Composite Artefacts in the Ancient Near East

Exhibiting an imaginative materiality, showing a genealogical nature

edited by

Silvana Di Paolo

Ancient Near Eastern Archaeology 3

Archaeopress Publishing Ltd
Summertown Pavilion
18-24 Middle Way
Summertown
Oxford OX2 7LG

www.archaeopress.com

ISBN 978 1 78491 853 8
ISBN 978 1 78491 854 5 (e-Pdf)

© Archaeopress and the authors 2018

Cover images: Details of a gaming board from Ur. Tomb PG 580. Early Dynastic IIIA. 2550-2400 BCE. Courtesy of Penn Museum, image #29557.

Cover design by Marco Arizza, Institute for Studies of Ancient Mediterranean - CNR.

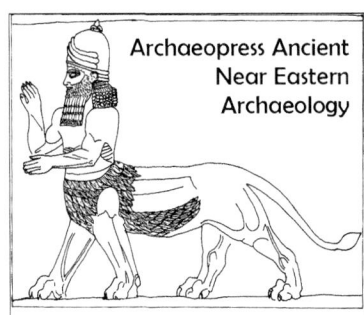

All rights reserved. No part of this book may be reproduced, or transmitted, in any form or by any means, electronic, mechanical, photocopying or otherwise, without the prior written permission of the copyright owners.

This book is available direct from Archaeopress or from our website www.archaeopress.com

Contents

List of Figures ii

Contributors v

Introduction: New Lines of Enquiry for Composite Artefacts? 1
Silvana Di Paolo

Section 1
The Planning: Materiality and Imagination

**From Hidden to Visible: Degrees of Material Construction of an 'Integrated Whole'
in the Ancient Near East** 7
Silvana Di Paolo

A Composite Look at the Composite Wall Decorations in the Early History of Mesopotamia 21
Alessandro Di Ludovico

Section 2
Symbols in Action

Composite animals in Mesopotamia as cultural symbols 31
Chikako E. Watanabe

Shining, Contrasting, Enchanting: Composite Artefacts from the Royal Tomb of Qaṭna 39
Elisa Roßberger

**Entangled Relations over Geographical and Gendered Space: Multi-Component Personal Ornaments
at Hasanlu** 51
Megan Cifarelli

Section 3
Sum of Fragments, Sum of Worlds

Composing Figural Traditions in the Mesopotamian Temple 65
Jean M. Evans

Polymaterism in Early Syrian Ebla 73
Frances Pinnock

**Near Eastern Materials, Near Eastern Techniques, Near Eastern Inspiration:
Colourful Jewellery from Prehistoric, Protohistoric and Archaic Cyprus** 85
Anna Paule

List of Figures

Section 1
The Planning: Materiality and Imagination

S. Di Paolo: From Hidden to Visible: Degrees of Material Construction of an 'Integrated Whole' in the Ancient Near East

Figure 1. Remains of a wooden column decorated with bronze bands on the floor of the entrance to the Shamash Temple 9
Figure 2. Gold applique probably intended for recycling from the Room R409, Stratum IIIc (Assyrian Trade Colonies) at Kaman-Kalehöyük 10
Figure 3. The creative process: materiality, memory and imagination..11
Figure 4. Fragment of a soapstone vessel from the Temple of Inanna, Nippur (Level VIIB) 12
Figure 5. Different approaches to the concept of imagination..13
Figure 6. Composite lion-headed aigle from Mari. 'Treasure of Ur'. Palace P-0. Third quarter of 3rd millennium BC..16
Figure 7. Detail of a gaming board from Ur. Tomb PG 580. Early Dynastic IIIA. 2550-2400 BCE.......17

A. Di Ludovico: A Composite Look at the Composite Wall Decorations in the Early History of Mesopotamia

Figure 1. Early Finding by Loftus..21
Figure 2. Left: Mosaics in situ at Tell 'Uqayr; top-right: Mosaics in situ by the Staircase Podium of the *Rundpfeilerhalle* at Uruk; bottom-right: *Bieberschwänze* and plaster of the *Steinstiftgebäude*............22
Figure 3. Digital Three-Dimensional Reconstruction of the *Pfeilerhalle* 23
Figure 4. Plan of the Northern Pillar of the *Pfeilerhalle* 24
Figure 5. Example of Analogic Sound Wave and its Digital Conversion 26

Section 2
Symbols in Action

E. Roßberger: Shining, Contrasting, Enchanting: Composite Artefacts from the Royal Tomb of Qaṭna

Figure 1a-b. a) Two drop-shaped pendants with variscite inlays, lapis lazuli bead between suspension loops only preserved in one case; b) rectangular object with lapis lazuli and asbestos-mineral inlays..........40
Figure 2. a) Rosette with carnelian and lapis lazuli inlays; b) small rosettes with carnelian inlays; c) reconstruction of rosette and gold-discs ensemble, all found in close distance; d) sphinx head from ʿAin Dara; e) terracotta plaque from Tell Munbāqa depicting nude female with rosette frontlet 41
Figure 3. a) Banded agate 'eye-stone' set in gold; b) 'double-eye' banded agate set in gold; c) two 'double-eye' jewels with glass inlays, interior filling not preserved; d) three roundels with concentric rings of carnelian inlays, filling material not preserved 42
Figure 4. a) Eye-like artefacts consisting of amber, lapis lazuli, variscite and gold strips; b) cross sections of two of the artefacts; c) schematic diagram of the human eye; d) find spots of eye-like amber artefacts at the Royal Tomb of Qatna 44

M. Cifarelli: Entangled Relations over Geographical and Gendered Space: Multi-Component Personal Ornaments at Hasanlu

Figure 1 Map showing location of Hasanlu...51
Figure 2. Site plan of the Hasanlu IVb Citadel, showing location of bead storage 52
Figure 3. Graph showing frequency of occurrence of artefact types in women's and men's and women's burials during Period IVb, ranked from left to right using a z-score calculation 53
Figure 4. Excavation photograph of Burial SK481, adult female, Operation VIF Burial 10 53
Figure 5. Composite photograph of HAS64-193 (UM 65-31-113), based on photographs of artefacts in situ 54

Figure 6. Beads, including *Arcularia*, from Middle Bronze Age Burial SK45-7, Hasanlu VIb, HAS 58-134 (UM59-4-78) 55
Figure 7. Object biography of beaded dress ornaments 56
Figure 8. Excavation photograph of Burial SK 448, adult female, Operation VIC Burial 4 57
Figure 9. Excavation drawing of Burial SK481, showing location of armor scale and arrangements of bead groups that accompanied it 58
Figure 10. Object biography of beaded composite ornaments featuring armor scales 59

Section 3
Sum of Fragments, Sum of Worlds

J.M. Evans: Composing Figural Traditions in the Mesopotamian Temple

Figure 1. Early Dynastic bull man 65
Figure 2. Tell Agrab, Shara Temple, Early Dynastic sculpture fragment of a male figure 66
Figure 3. Nippur, Inanna Temple VIIB, Early Dynastic statue of a standing female figure 67
Figure 4. Nippur, Inanna Temple VIIB, Early Dynastic statue of a standing female figure 68
Figure 5. Tell Asmar, Abu Temple, Early Dynastic statue of a standing male figure 68
Figure 6. Khafajah, Small Temple, Early Dynastic pendant of a human head 68
Figure 7. Old Babylonian baked clay sculpture of a female head 69
Figure 8. Tell Agrab, Shara Temple, Early Dynastic pendant of a human foot with scorpion incised on sole of the foot 69
Figure 9. Al-Hiba, Area B, baked clay foot 69
Figure 10. Nippur, Area WA, Kassite clay figurine 70

F. Pinnock: Polymaterism in Early Syrian Ebla

Figure 1. Ebla, schematic plan of the Royal Palace G 74
Figure 2. Ebla, arm-rest, probably from a throne, wood and shell, c. 2300 BC, from the Royal Palace G, L.2601 75
Figure 3. Ebla, wooden plank with two carved male heads, probably from a cupboard door, wood, from the Royal Palace G, L.2764 75
Figure 4. Ebla, revetment of a foot, gold, ca. 2300 BC, Royal Palace G 75
Figure 5a. Ebla, the steps of the Ceremonial Staircase, ca. 2300 BC, Royal Palace G 76
Figure 5b. Ebla, entrance to the Administrative Quarter, with limestone threshold and wood and shell decoration, ca. 2300 BC, Royal Palace G, L.2875 76
Figure 6. Ebla, three views of a segment of female hair-dress, steatite, ca. 2300 BC, Royal Palace G, L.2752, on the last step of the Monumental Stairway 76
Figure 7a. Ebla, figure of standing leopard from a wall inlaid panel, limestone with lost inlays of different materials, ca. 2300, Royal Palace G, L.2913 77
Figure 7b. Ebla, figure of passing human-headed bull from an inlaid wall panel, limestone with lost inlay for the beard, ca. 2300 BC, Royal Palace G, L.2913 77
Figure 8. Ebla, figure of standing prisoner, limestone with lost inlays for the rope holding his arms and for the background, ca. 2300 BC, Royal Palace G, L.2913 77
Figure 9. Ebla, fragments of bridles from wall inlays, lapis lazuli, ca. 2300 BC, Royal Palace G 77
Figure 10a. Ebla, reconstruction of a procession of officials from an inlaid wall panel, limestone and lapis lazuli, ca. 2300 BC, Royal Palace G 77
Figure 10b. Ebla, reconstruction of three front facing kings' figures from inlaid wall panels, limestone, ca. 2300 BC, Royal Palace G 77
Figure 11. Ebla, three rosettes from the decoration of inlaid wall panels, limestone, ca. 2300 BC, Royal Palace G 78
Figure 12. Ebla, three views of a composite female hair-dress, steatite, ca. 2300 BC, Royal Palace G, L.2862, at one side of the entrance to the Throne Room L.2866 78
Figure 13. Ebla, fragments of a textile, or of the decoration of a textile, gold, ca. 2300 BC, Royal Palace G. L.8778 78
Figure 14. Ebla, miniature figure of human-headed bull, possibly from a standard, gold and steatite, ca. 2300 BC, Royal Palace G, L.2764 79
Figure 15. Ebla, miniature figure of veiled woman probably from a standard, limestone, steatite and jasper, ca. 2300 BC, Royal Palace G, L.3600 79
Figure 16. Ebla, reconstructive drawing of the *maliktum*'s standard, wood, steatite, limestone, gold, silver and jasper, ca. 2300 BC, Royal Palace G, L.9330 79

Figure 17. Ebla, miniature head of lion on a mobile support, wood and gold, ca. 2300 BC, Royal Palace G, L.2984 80
Figure 18a. Ebla, wooden core of Tabur-Damu's figure from the *maliktum*'s standard, showing the complete modelling of the face, ca. 2300 BC, Royal Palace G, L.9330 80
Figure 18b. Ebla, arm of Tabur-Damu's figure from the *maliktum*'s standard, showing the modelled wooden core and the thick silver coating, ca. 2300 BC, Royal Palace G, L.9330 80
Figure 19a. Ebla, back part of the hair-dress of Tabur-Damu's figure from the *maliktum*'s standard, steatite, ca. 2300 BC, Royal Palace G, L.9330 81
Figure 19b. Ebla, lower edge of the hair-dress of Tabur-Damu's figure from the *maliktum*'s standard, steatite, ca. 2300 BC, Royal Palace G, L.9330 81
Figure 20. Ebla, drawing of the separate pieces composing Tabur-Damu's figure from the *maliktum*'s standard, wood, steatite, silver and jasper, ca. 2300 BC, Royal Palace G, L.9330 81
Figure 21. Ebla, dress of Dusigu's figure from the *maliktum*'s standard, gold, ca. 2300 BC, Royal Palace G, L.9330 81
Figure 22. Ebla, drawing of the separate pieces composing Dusigu's figure from the *maliktum*'s standard, wood, steatite, gold, marble and jasper, ca. 2300 BC, Royal Palace G, L.9330 82
Figure 23. Ebla, proposal of reconstruction of the en's standard, steatite, gold, limestone, red stone and shell, ca. 2300 BC, Royal Palace G, L.2982 83

A. Paule: Near Eastern Materials, Near Eastern Techniques, Near Eastern Inspiration: Colourful Jewellery from Prehistoric, Protohistoric and Archaic Cyprus

Figure 1. Loop pin from Enkomi Br. T. 19 86
Figure 2. Loop pin from Enkomi Br. T. 19 87
Figure 3. Composite artefact (pin?) from Enkomi O.T. 74 89
Figure 4. Ivory pomegranate from Enkomi (uncertain tomb) 89
Figure 5. Necklace from the sanctuary at Arsos (detail) 91
Figure 6. Broad collar (*usekh* collar) from Enkomi Br. T. 93 91
Figures 7-8. Finger ring from Kouklia-Evreti T. VIII (upper surface/profile) 92

Contributors

Megan Cifarelli is a Professor and Chair of the Department of Visual Studies and Art History at Manhattanville College in Purchase, NY. Her main areas of research are theoretical and methodological approaches to the study of representations and instances of dress in ancient contexts, focusing in particular on Assyria, northwestern Iran and the Caucasus in the early first millennium BCE. She is the author of multiple articles dealing with the material culture of Hasanlu, Iran, theoretical issues relating to mixed material culture produced in contact situations, and Assyrian visual culture. Her recent co-edited volume (with Laura Gawlinski) on dress, What Shall I Say of Clothes, was published by the Archaeological Institute of America in 2017.

Alessandro Di Ludovico is Doctor in Near Eastern archaeology and art history. His research activity is mainly focused on visual languages and cultural history of ancient Mesopotamia and Syria, as well as on digital humanities (especially the use of quantitative methods in the investigation of ancient visual languages). On the field, he participated in numerous excavation and campaigns in Syria and Palestine. Since 2014 he cooperates with research projects on historical geography which are still being developed at Sapienza University of Rome. He took part in several international meetings and congresses, sometimes in the role of or co-organiser of whole symposia or single sessions.

Silvana Di Paolo (PhD Rome 2001) is, since 2001, researcher at the Institute for Studies of Ancient Mediterranean of the Italian National Council of Research (CNR). She is the Director of the Series Biblioteca di Antichità Cipriote, Scientific Board Member of al-Sharq (published in Paris) and Editorial Board Member of Rivista di Studi Fenici published by ISMA. As CNR researcher she is co-coordinator of different projects in collaboration with European and not-European foreign institutions. Actually she is Co-Director of the QaNaTES project in the Iranian Kurdistan. She has written extensively on the relationship between art and power, location and styles of workshops, social meaning of works of art, as well as on material culture of the 2nd millennium BC She is currently working on the concepts of similarity in assemblages of artifacts and routinisation of the artisanal production in the ANE, as well as on the applications of the Shape and Semantic Analysis on Mesopotamian glyptics.

Jean M. Evans is the Chief Curator and Deputy Director of the Oriental Institute Museum and a Research Associate of the Oriental Institute. She was previously a fellow in the Munich Graduate School for Ancient Studies at Ludwig Maximilian University and has been the recipient of fellowships from the Getty Foundation, the American Academic Research Institute of Iraq, the Warburg Institute, and the German Archaeological Institute. Jean was a curator at the Metropolitan Museum of Art from 1999-2008 and was ultimately a co-organizer of the exhibition Beyond Babylon: Art, Trade, and Diplomacy in the Second Millennium BC and co-editor of its corresponding publication. She is also the author of The Lives of Sumerian Sculpture: An Archaeology of the Early Dynastic Temple (Cambridge University Press, 2012).

Anna Paule is an Austrian Doctor of Philosophy in Archaeology. After her Master's Degree in Classical Archaeology at the University of Salzburg, Austria, she moved to France where she received her PhD at the Université Aix-Marseille I, Aix-en-Provence, in 2013. After six years of archaeological studies in France and several research stays in Greece, Cyprus, and England she moved back to Austria; since then she has been an independent researcher in Linz. Her research interests are on the Bronze Age and Iron Age Mediterranean, or, more precisely, on the role that Cyprus played during the transition from the Late Bronze Age to the Early Iron Age; on metallurgy, jewellery, and on further small objects indicating cross-cultural contacts.

Frances Pinnock (Rome 1950) is Associate Professor of Archaeology and Art History of the Ancient Near East in the Sapienza University of Rome, and is co-Director, with P. Matthiae, of the Italian Archaeological Expedition to Ebla, of which she is a member since 1971. She is author of six scientific monographs and of more than 90 articles in scientific journals. Her main interests are the archaeology and history of art of pre-classical Syria, the transmission of iconographies and the roles of women in the ancient Near East.

Elisa Roßberger is a postdoctoral fellow at the Munich Graduate School for Ancient Studies 'Distant Worlds'. She received her Ph.D. in Near Eastern Archaeology from Eberhard-Karls-University Tübingen with a thesis on jewellery objects discovered at the Royal Tomb at Qatna (published in 2015). She taught seminars and conducted research at the Institutes of Near Eastern Archaeology at Freiburg and Munich, and excavated at various sites in Syria, Lebanon and Iraq. Her scientific work aims at a combined analysis of archaeological and textual sources, with a current focus on changes in the materiality and mediality of artworks in third to early second millennium BCE Mesopotamia.

Chikako E. Watanabe (PhD Cambridge 1999) is Associate Professor of Art History in the Faculty of International Studies at Osaka Gakuin University. Her academic interests range from Neo-Assyrian pictorial narratives and animal symbolism to an analysis of the source materials of Assyrian reliefs and cuneiform tablets. She was awarded the Third JSPS prize on 'Narratological Interpretation of the Art of Ancient Mesopotamia' in 2006. She is the author of *Animal Symbolism in Mesopotamia: A Contextual Approach*, WOO 1 (2002).

Introduction.
New Lines of Enquiry for Composite Artefacts?

Silvana Di Paolo

Enrico Prampolini was the principal theoretician of futurist polymateric art. He was experimental in terms of technique and materials, testing a number of different methods and using a combination of materials. In 1915, when establishing the theoretical foundations of what he called 'plastic complexes' he had said that: '*l'arte polimaterica non è una tecnica ma – come la pittura e la scultura – un mezzo di espressione artistica rudimentale, elementare, il cui potere evocativo è affidato all'orchestrazione plastica della materia...Nel* polimaterico, infatti, il valore evocativo si manifesta inversamente alla reazione visiva esterna, poiché opera nelle ragioni irrazionali dello spirito. Introspezioni che – nella sfera delle arti plastiche – creano dei sistemi, *delle costanti di* reazioni interne, *le quali producono a loro volta* – successivamente *e simultaneamente* – *i fenomeni della* meraviglia, *della* sorpresa, *del* miracolismo spettacolare. *Da questa magia della* materia *nelle sue* apparizioni bioplastiche, nasce il nuovo incantesimo dell'*arte polimaterica,* del *polimaterico'*.

If the macroscopic result of such a definition for the futurist polymateric art is that each component has a role within larger assemblages in physically building the artwork by developing his expressive possibilities and sensuous qualities (three-dimensional, tactile or volumetric effects), then different points of view highlighted how the true aim of this art was not the representation of the exterior and sensory reality of the world, but the creation of 'spiritualising machines'. This term refers to the possibility of building a heterogeneous universe and to inject reality into a work of art, so as to achieve an absolute realism intended as simultaneity of states of consciousness. Thus, the polymateric compositions' power is the power to 'spiritualise' matter by the subject observing it. The matter always stays inert, dead, inexpressive, if it is not led to spiritualise itself. Such artistic experiences, promoted by the futurist's movement exalting psychic energy, permit us to observe reality from a hyperreal point of view, as well as to recreate reality through a new, spiritual mode of artistic creation.

The implications of these approaches to evaluating the creation of 'polymateric plastic' at the beginning of the 20th century and his meaning, from this very different perspective well summarise the opposite points of view giving impetus to the discussion on ancient polymateric artefacts and their qualities in antiquity, and in particular with reference to the ancient Near East. On the one hand are the objective and natural attributes of materials, possibly exalted from their transformation: a form of fascination immanent in all kind of technical activity which promotes the transition from the ordinary into a different realm, imbuing the object with new meaning. On the other hand is the idea that properties of materials are not fixed attributes of 'matters', but are processual as well as relational: the qualities of artefacts are subjective and included in the worldview of artisans making them, as well as in the mind of who observes or appreciate them.

Taking into account these premises and theoretical tenets, the complex relationship between environment, materials, society and materiality, with particular reference to the composite artefacts in the ancient Near East, is the topic of the eight papers gathered together in this volume. They were first presented as part of an International Workshop organised by myself and held in Vienna at the Austrian Academy of Sciences during the 10th International Congress on the Archaeology of the Ancient Near East, 26th April 2016. This inquiry is, obviously, based on whole or incomplete artefacts found during excavations, although they represent just a small percentage of the entire realm of archaeological material, and of myriad combinations possible therein. Some categories of artefacts certain to have been used in antiquity, have completely disappeared or misunderstood in the excavation reports: for instance, metal and/or stone components applied to textiles, or combinations of leather and wood (both completely disintegrated and lost) etc. It is therefore impossible, by physical material remains alone, to accurately know all of the possible combinations of past material culture. Some of these categories of lost artefacts are mentioned or described in cuneiform texts. At Mari, luxurious textiles used to make royal garments were probably richly decorated with pieces of blue stone or metal discs, although these are now archaeologically invisible. This also demonstrates the difficulty in connecting textual evidence and material remains: not only is taphonomic loss an issue, but the variation in shapes and combinations of composite artefacts also make the correlation more complex.

Starting with material remains, we can investigate some aspects correlated to specific characteristics of the artefacts. Firstly, the method to describe, organise or standardise them is strictly linked to the adopted terminology. In modern language, as well as in contributions gathered in this volume different terms are used to indicate these types of objects (composite, multi-component, polymateric etc.) but not all are unambiguous. This was one of the first issues that I faced when organising this workshop.

The adjective 'composite', for instance, introduces a first point for discussion. The Collins Dictionary defines 'composite' as 'composed of separate parts', while the Oxford Dictionary listed two different definitions: 1) 'made up of several parts or elements'; and 2) (of a constructional material) made up of recognisable constituents'. Certain materials, like bronze or glass, are composite: constituent elements with different physical and chemical properties which, when combined, produce a new material, with characteristics different from the individual components that remain separate and distinct within the finished structure. In such cases, the new composite material is preferred for many reasons: because it is stronger, or more sophisticated and, sometimes, less expensive when compared to traditional materials. Ceramics, for instance, may be considered the first manufactured composite material. The production process involves several stages, changing the appearance and properties of the raw materials: natural clay is, in fact, a mixture of clay minerals and 'accessory' minerals, derived from sediments or rocks, adding other substances. Otherwise, our field of study concerns the composite objects produced or shaped by human craft, the result of a combination of elements of distinct material and colour, but which preserves, their physical and chemical properties. They are a sum of their parts or fragments: each element possesses a specific material nature and unique origin.

But the evidence is more complicated.

The silver cup inlaid with bucrania from Late Bronze II Enkomi (Cyprus) has been discussed, for a long time, as containing niello (a composite metal alloy fused onto base metal for decorative effect), as do some bronze daggers found in the Shaft Graves at Mycenae. However, analyses carried out on these artefacts reopened the question, because the dark material visible on the surface is, without any doubt, blackened bronze: the darkness derives from a thin surface patina created after immersion in a chemical bath. In this case, the metal technology is tailored to particular needs. On the one hand, there is a composite material, that is the blackened bronze, formed by the combination of more chemical elements and then modified through a deposition of a thin patina. On the other hand, as composite artefacts, these daggers are characterised by a black-inlaid bronze. It is inserted in a conventional bronze frame and becomes, in turn, the dark support for exalting colour and silhouettes of a series of gold and silver inlays in which figurative subjects predominate.

Thus, the craftsmanship is oriented to the achievement of sophisticated products through assemblage techniques and the blending of contrasting properties and qualities of materials. Here, the term 'composite' is a combination of the power of technology and the ability to form new images: the strict relationship between creativity, technology and manufacture produces novel interactions and solutions. Therefore, scientific analyses have become fundamental tools to explore the adopted technological solutions; from the nature and origin of materials to their workmanship, assembling techniques and fixing procedures. This dataset would allow us to establish, where possible, an appropriate terminology and an easier descriptive system.

Composite artefacts can be considered the result of a special connection between human brain/body and environment. At least, since the Upper Palaeolithic, they form a point of interaction. On the one hand, the material sphere (the external world) where materials, exist in nature and modified by humans, are adapted, transformed, and assembled to produce a finished object. On the other hand, there is the cognitive-cultural sphere, which is engaged in performing constructive actions. Understanding the fundamental architecture of human cognitive processing, especially how it interacts with cultural contexts and manifests in the production of composite artefacts, requires significant further research. In fact, in the present volume, all authors just preliminarily approach this aspect (especially in the contributions by myself and A. Di Ludovico).

Materials and their natural properties, including colours (modified and enhanced through manufacturing processes) are used for promoting and extolling virtues and qualities. Artisans amplify the composite nature of the artefact by assembling garments, body parts, and landscape in a uniquely imaginary world: contrasts of colours and the interplay of transparency or opacity illuminate the materials around them, exemplify the value of polychromy. However, 'polychrome' is a separate colour concept: in the ancient Mesopotamian texts, this term, focus on the ideas of 'variegation' and 'patterning' (for instance, it is applied to embroidered textiles), and brings together all of the meanings of colours, emphasising the congruity of parts to their whole, as well as an higher value for the multiplicity.

A primary concern of this volume is to provide specific case studies in which theoretical assumptions and hypotheses can be applied to the ancient evidence.

For this purpose, most of the papers take not only the general perspective, such as the relationship between materials and humans, but also a defined body of evidence – material, textual, visual, architectural - through which to address the issue.

The volume begins with a section entitled 'The Planning: Materiality and Imagination' which provides some remarks on the relationship between imagination and skill: they have to work together with sets of beliefs and myths explaining the origins of the world, as well as the symbolism and ideology of power in the construction of the composite artefacts. I myself examine this issue. With particular reference to the network of associations for the construction of composite 'bodies', I emphasise the synthesis between thought and images, between 'externalised' reality and 'internalised' man. Alessandro Di Ludovico's article analyses the organisation and meaning of the assemblages of cones used to create wall decorations in the Lower Mesopotamia during the first urban phase. The planning of such mosaics, made by different stones, as well as by clay specimens coloured on their ends or differently baked in order to nuance pieces, addresses the need of a depersonalised coding system aimed to the communication and interpersonal relations between communities.

The second section, 'Symbols in Action', consists of three articles investigating the symbolic dimension of composite artefacts. As signs creating associations, they, in some cases, offer a means to enchain relations between peoples, things and places. At the same time, semiotic introduces the social dimension of meaning, as well as the social processes of signification and interpretation. Chikako Watanabe's study introduces the relationship between texts and images. Focusing on composite animals - the lion-headed eagle and the lion dragon - exhibiting a body structure that consists of multiple body parts taken from different animals from reality, the author examines some aspects from the point of view of the aetiological and symbolic functions, as well as their relation to the materiality through some composite artefact conveying the same values and notions. The next article explores the concept of enchantment. Taking into account the inventories of the tombs discovered under the Royal Palace of Qatna (Syria), Elisa Roßberger emphasises the importance of the visual and semantic qualities of materials and colours for the artistic productions of Syria in the 2nd millennium BC. Composite artefacts are planned with a social function: in this case, enchantment, founded on the exceptional assemblage of materials, produces 'dramatic' and profound effects on viewers. Special personal ornaments discovered inside the female burials of Hasanlu (northwestern Iran) are the subject of the study proposed by Megan Cifarelli. They consist of iron armor scales with attached garments pins, stone, shell and composite beads. The symbolic dimension is particularly evident here: the author suggests that the creation and recycling of these aggregates of objects, which include fragments of masculine armor and emanate evocative sounds, enchain individuals across gendered boundaries.

The last section 'Sum of Fragments, Sum of Worlds' presents some case studies exploring the essence of composite artefacts as a sum of fragments but serving for the whole. The articles here included investigate how, although each element possesses a singular material nature, origin and context, aggregates of matters can inviting novel interactions in the material world from the social and religious points of view. Jean M. Evans analyses the process of building up of the body in the Early Dynastic temple sculpture. Although there was a plan for drilling the holes in order to assemble perfectly the different parts and create a body unity, she stresses that the construction of stone human or animal images combining fragments from different materials emphasises the meaning of the single body parts, while the role of the material properties (type, provenance, availability etc.) looks little relevant. As in Early Dynastic Mesopotamia, polymaterism is also a diffused practice in Northern Syria. Frances Pinnock examines the remains of composite artefacts coming from different areas of the Palace G (Early Syrian period). The palace furniture, found in scattered pieces, as a consequence of the dramatic fall of the town, includes composite objects made of many different materials (wood of many qualities, stone, gold). In particular, the *maliktum*'s standard represents an extraordinary synthesis of substances, each of them exalting specific divine and royal attributes. With the last contribution, the focus shifts to the island of Cyprus in a long period between the Late Bronze Age and the Archaic period. Normally, jewellery consists of many decorative items worn for personal adornment and made of different materials (metals, stone, organic materials etc.). Anna Paule investigates the goldsmithing traditions, stressing out how the wide range of materials employed emphasise the specific qualities of single and rare elements through the color combinations and composition patterns.

This volume represents a first attempt to conceptualise the construction and use of composite artefacts: the richness of approaches, the development of new issues depending on specific case studies, and the overturning of widely accepted ideas show the interest towards this category of objects and the opportunity to enlarge this field study in the future.

I would like to thank all the participants for the wonderful and stimulating Workshop. Many thanks also to the Vienna Organising Committee of the 10th ICAANE for allowing us to address this topic.

Last, but not least, I would like to express my appreciation and gratitude to Marco Arizza (ISMA) who, with patience and care, created the book cover design for this volume.

References

Chessa, L. 2012. *Luigi Russolo, Futurist: Noise, Visual Arts, and the Occult*. Berkeley and Los Angeles: California University Press.

Di Paolo, S. 2016. Beyond Design and Style: Enhancing the Material Dimension of Artefacts through Technological Complexity, in C. Suter (ed.), *Proceedings of the Workshop 'Levantine Ivories of the Iron Age: Production, Consumption, and Style'. 61st Rencontre Assyriologique Internationale, Geneva-Bern, 22-26 June 2015: Altorientalische Forschungen 42/1*: 71-79.

Knappett, C. 2014. Materiality in Archaeological Theory. In C. Smith (ed.), *Encyclopedia of Global Archaeology*: 4700-4708. New York: Springer.

Riccio, G. and F. Pirani (eds) 2016. *Laboratorio Prampolini. Disegni, schizzi, bozzetti, progetti e carte oltre il Futurismo, MACRO Museo d'Arte Contemporanea Rome, November 10, 2016 – January 15, 2017*. Rome: Silvana Editoriale.

Thomas, N.R. 2011. Recognizing Niello. Three Aegean Daggers, in A. Vianello (ed.), *Exotica in the Prehistoric Mediterranean*: 146-161. Oxford: Oxbow.

Thomas, A. 2015. In Search of Lost Costumes. On Royal Attire in Ancient Mesopotamia, with Special Reference to the Amorite Kingdom of Mari, in M. Harlow – C. Michel and M.-L. Nosch (eds), *Prehistoric, Ancient Near Eastern and Aegean Textiles and Dress. An Interdisciplinary Anthology*. (Ancient Textiles Series 18): 74-96. Oxford and Philadelphia: Oxbow.

Section 1

The Planning: Materiality and Imagination

From Hidden to Visible.
Degrees of Material Construction of an 'Integrated Whole' in the Ancient Near East

Silvana Di Paolo
Consiglio Nazionale delle Ricerche – Istituto di Studi sul Mediterraneo Antico
silvana.dipaolo@isma.cnr.it

Abstract

Artefacts made of different materials are, by their nature, a particular way of interacting with the world, and an interface between the material sphere, where raw materials are physically transformed and assembled, and the cognitive sphere, where different areas of the brain are involved in performing hierarchical constructive actions: therefore, they possess a transformative agency in relation to their creation and use. In the ancient Near East, the evidence for the widespread use of this kind of artefacts (statuary and small sculptures, jewellery, vases, furniture fittings, etc.) is undisputable. As aggregates of associations and relationships, they can assume different meanings according to their 'construction' and assemblage. This study investigates some preliminary issues, such as terminology and categorisation according to the archaeological remains: they are necessary tenets for understanding use and meaning of these objects. Besides, an examination of different categories of polymateric artefacts until now classified (by superimposition, insertion and juxtaposition) identifies them as nodes in a web of connections between creativity, mental planning, and making, in this way permitting to explore the complexity of their manufacture (from the choice and availability of materials to the assemblage of contrasting material properties and colours).

Keywords: Polymateric, categorisation, superimposition, insertion, juxtaposition, materiality, imagination, metaphor

1. The Potential of the Material.
Construction Processes to 'Sature' Artefacts with Meaning

When one approaches the study of the artefacts made of different materials coming from the stratified archaeological contexts of the ancient Near East,[1] one cannot overlook some preliminary aspects strictly linked to function and technology of these items.

As regards the nomenclature, some remarks were already made in the Introduction, where I emphasised the ambiguous use of the term 'composite', not only because it is open to interpretations denoting either composite artefacts (produced with aesthetic aims) or composite materials (produced with practical aims), but also because the material evidence shows cases of overlapping: technology used with artistic goals, polymateric items manufactured with practical objectives (see Introduction). In the descriptive process of these particular items, the typological analysis is probably a useful mean of comparing artefacts from different sites and methods of manufacture, while as regards their function it is impossible to define assured categories, because the inquiry is often based on incomplete artefacts.

Although the categorisation process currently being carried out by the author is still a work in progress, it is possible to identify many classes of polymateric artefacts in the ancient Near East, according to the ways of combination of materials and the methods of manufacture. This evidence has to be compared with contemporary texts that, in many cases, refer to specific classes of objects (divine statues and temple furniture, for instance), as well as to materials combined together. A classification exclusively based on the materials used is, instead, a serious problem, because of the incomplete or extremely fragmentary state of archaeological findings.

In some cases, polymateric items (or what is preserved of them) are characterised by the presence of multiple materials but one or more of them is, as a rule, hidden, or partially 'obscured'. In this **superimposition** process one (or more) of the materials is overlaid and used as a second skin, thus becoming indistinguishable from the underlying form of the work.

Once completed, the object, commonly, can also assume a mimetic and illusionistic aspect which transforms the base material into an 'indifferent' material.[2]

---

[1] A volume entitled *Divinely Inspired, Divinely Planned! Approaching Multi-Materiality in the Ancient Near East* is in preparation.

[2] Although this same principle could be applied to painted artefacts, they are different and are not included here. For a general and useful discussion of the polychromy of different media (marble, stone, bronze, wood etc.) in antiquity, see Policromia 2004.

A macroscopic use of polymateric artefacts by superimposition is known in the ancient Near East, at least, from the 4th millennium BCE, although the material culture and texts describe the same kind of artefact in different ways. What is obscured (at least partially) is, often, a less noble material: wood, for instance, was covered and overlaid with precious metal sheets (gold, silver, bronze) but also less precious metals, such as copper or bronze, were used as 'obscured' materials (cf. infra). Examples of massive wooden gates overlaid with bronze are known at Balawat, Nimrud, Khorsabad and Ashur. At Balawat, three pairs of gates (standing to a height between 6-8m) have been found, two dating from the reign of Ashurnasirpal II and one from that of Shalmaneser III. According to the reconstruction hypothesised by H. Güterbock, the bronze bands would have been arranged in groups, leaving a wide gap in the middle of the doors[3] where wood was largely visible. The strips were probably arranged so that the best-executed and more easily visible were placed near the bottom of the gates, while more crudely decorated sheets of metal were put at the top.[4]

The royal Assyrian inscriptions confirm the existence, archaeologically proven, of polymateric artefacts formed by wood and metal assembled together. The overlaying of wooden gates with many metal bands was a common practice from the Middle Assyrian period onwards. Whether references concerning wood types are sometimes lacking, the record of overlaid materials is a common practice. Metal bands are described as simple sheets of bronze or copper fixed on wooden gates or as probably alternate decorative sheets of silver and copper (for the 'Palace without Rival' of Sennacherib), or gold and silver (for the Temple of Esagila at Babylon by Esarhaddon), always applied onto the less noble material.[5] The alternating metals/colours represent the dominant feature and the main goal of the polymateric gates. The apparently more simplistic use of bands of copper or bronze, although enhancing less 'value' to the artefacts, is a result of a regulated activity.

Recently, it was observed, in fact, that considerable effort was put into creating copper alloys for specific purposes. The Assyrian bronze composition shows that added tin and lead contents were carefully controlled in order to produce metal of the desired colour.[6] If the amount of tin is less 5% the metal melts at a lower temperature and is easier to cast, but the final colour is not far from the reddish hue of pure copper. But if the amount of tin is c. 10% (as in the Assyrian bronze composition) and the alloy include lead, zinc, silver, iron and so on, both the casting technology (for the presence of lead) and final colour produce a better product and a different effect. Perhaps this result is called 'burnished copper' in the contemporary texts.[7] A different use of polymateric artefacts by superimposition is again attested in the Neo-Assyrian period. The entrance of the Sin Temple at Khorsabad was flanked by two columns made of cedarwood (9m high) and overlaid with bronze and gold. Sections of bronze and gold sheets decorated with an embossed scale-like design encased the wood shafts and were fixed to them with metal nails. According to the excavator, Victor Place, the decoration of these metal overlays seemed to imitate the cortex of trees.[8] The manufacturing technique is not well known. Perhaps bronze or gold sheets were adhered to a mould and cold hammered on it; but it is also possible that pointed tools were used to create the scale design and, then, the metal bands were hammered from the reverse side in order to obtain the final effect in relief. Whatever the technique was used, the combination wood-metal had different purposes, among which was the possibility to economise (precious) metal, more easily produce ornamentation, enhance the value of columns, and ennoble the temple building. Similar bronze bands decorated at least a wooden column flanking the entrance of the Shamash Temple on the right side, again at Khorsabad. These large sheets (70cm wide) each decorated with two series of figures divided by rosettes were affixed to the wooden column, probably by nails.[9] The original assemblage is clearly visible in an excavation photo of the American mission (Figure 1).

The shine of gold, its apparently indestructible nature, its malleability, and its relative scarcity made it an ideal material to embody divine qualities.[10] In fact, its specific physical properties make it one of the main 'ostentatious' materials and the perfect means of transmission of notions perceived as invariable, and, consequently, transcendental.[11] Because gold is perceived as an appropriate material with which to address the gods, temples and palaces were decorated with golden and gilded images and statues. In particular, the polymaterism realised through the superimposition can be interpreted, in some cases, as an overlapping of symbolic meanings and cultural values. The combination metal-metal seems to be emblematic in some specific cases.

The production of polymateric divine and royal statuettes is well known in the Bronze Age Levant. Recent analyses carried out on some metal anthropomorphic figurines representing the local gods (Ba'al, El) discovered at Ugarit and Minet el-Beida[12]

---

[3] Güterbock 1957: pl. 22.
[4] Curtis 2013: 58-61.
[5] Luckenbill 1924: 106, ll. 27-28; Luckenbill 1926-7: vol. II, 247, § 653.
[6] Ponting 2013: 220-242.
[7] Luckenbill 1924: 140-141, lines 5-16.
[8] Loud 1936: 97-98.
[9] Loud 1936: 104-105; Curtis 2013: 61.
[10] Winter 2012.
[11] See the recent remarks by Meller, Risch and Pernicka 2014: 11-17.
[12] Cluzan 2007: 71-89.

Figure 1. Remains of a wooden column decorated with bronze bands (indicated as a and b) on the floor of the entrance to the Shamash Temple. (Loud 1936: 104 fig. 111).

have revealed the use of gold for tiaras and faces, whereas divin e bodies and clothing were completely covered by a white metal (tin or silver), according an Anatolian[13] or Egyptian tradition.[14] Although it could be also possible that the stratum of tin/silver was hidden, covered by gold applied on all the surface of the body of Ba'al,[15] nevertheless there are no traces of tin under the tiara, at least on one case.[16] Thus, the emphasis was, probably, on the assemblage of different materials, textures and colors. The shine and 'reflectivity power' of the white metal was contrasting with the yellow-red gold amplified by the color of a reddish stratum applied under the metal.[17] The colours and an historically contingent symbolism of the materials were enhanced by the use of wood or stone[18] for the throne of El. The bright and divine golden body was, perhaps, associated with other self-evident properties: the permanence and stability of stone or, conversely, the transience of wood, a more 'human' and trasformative element.

A bronze statuette (30cm high) discovered at Hazor in Northern Israel probably represents a local enthroned king and is dated to the mid-18th or 17th century BCE. The statue remained in use for four centuries until it was buried prior to the destruction of the city. Traces of gold survive on different body parts: headgear, torso, arms etc. Thus, the figure was completely covered by gold leaves. This aspect is probably linked to the technique used to coat this statue with gold, as demonstrated by Tallay Ornan. The separately cast right arm and left forearms were also encircled by pieces of gold at the point where they were inserted in the body. Thus, less visible parts also were gilded.[19] The ways in which an individual craftsman confronted the materials in the manufacture of particular objects depends on the variety of technical solutions that arise in response to a given objective. But a given technical solution is not always governed solely by the state of the technology itself. Techniques are as much a part of the aesthetics, the religion, the economics, the utilitarian objectives of the social and cultural milieu: it becomes part of a specific 'point of view'.[20] My suggestion is that the external application of thin sheets of gold leaves has to 'obscure' the underlying metal and to completely envelop and enclose, without any interruption, the royal body and his attributes (the headgear, for instance) in order not only to transfer eternal qualities to it, because gold maintains its physical existence eternally, but also to represent the eternal character of the relationship between the image/king and worshippers/subjects. Thus, this technical act becomes a form of containment that can be used to conceal/transform, reveal/exhibit, and conserve/preserve the royal body.[21] Gilding is put in contact with his content, becoming a boundary to create an interface between subject and the world.

---

[13] It is my opinion based on Hittite texts. See p. 17-18.
[14] Cluzan 2017: 85.
[15] Cluzan 2017: 85, note 57.
[16] Cluzan 2017: 84.
[17] Cluzan 2017: 77.
[18] Cluzan 2017: 86 who postulates the use of a *matière minerale* because the restauration techniques have not identified traces of wood.
[19] Ornan 2007: 445-458.
[20] On this aspect, see Lachtmann 1971: 26.
[21] On this concept, see Douny and Harris 2014: 15-18.

In the ancient Near East also the technique of the **insertion** is well attested. This *modus operandi* is characterised by a principal matter (metal, stone, etc.) and the addition of other materials with a predominantly ornamental function set into the surface/structure of it, without compromising the basic design of the work. As with the preceding category, this typology of objects features simultaneous visibility of various materials. Here is included goldsmith work, often characterised by the use of 'secondary' materials (semi-precious stones). As regards this category of artefact, it is necessary to pay attention to the organisation of jewellery items, because their original arrangement has often largely been lost. In many cases, in fact, excavators reconstruct complete and polymateric 'necklaces' without explaining the reasons for such reconstructions.[22]

The evidence shows that, at least, this system of mixing media include the reuse of 'spoliate' materials, although this is also possible for other categories of artefacts.[23] Some recently discovered artefacts give some indications about the hypothesis of remelting for successive reuse. One such was unearthed from Kaman-Kalehöyük (central Anatolia): stratigraphically assigned to the Stratum IIIc, it is dated to the Middle Bronze age. (Figure 2).[24] It could be considered an 'ecological' object, subject to an extended phase of use and recycling (small insertions of different materials originating by other artefacts, waste material or remelted metal); changes in social and cultural environments could act by inviting novel interactions with new materials and technologies.

This fragmentary gold artefact probably represents a hybrid creature, consisting of a winged lion-dragon in profile to left rearing on its hind legs and originally associated to another mirror specimen for a specular composition. Probably introduced into Anatolia during this period, this Mesopotamian figure attests to the participation of this site along the trade route between Anatolia and Assyria during the first half of the 2nd millennium BC.[25] It is constructed from a hammered gold sheet cut to form the figure. On one side, it is enhanced by a series of circular gold cloisons (discovered empty), while on the other side the piece is undecorated.[26] This animal silhouette, delimited by gold wires, was perhaps enriched with multicoloured (semi-precious) stones or other materials. They were also inserted into the circular cloisons applied along the animal body and the base.[27] The object, perhaps together with another specular specimen, was mounted on a different support

Figure 2. Gold applique probably intended for recycling from the Room R409, Stratum IIIc (Assyrian Trade Colonies) at Kaman-Kalehöyük (Omura 2015: 11, Fig. 1).

(wood or other): mounting holes are still visible on the base under the paws of the lion-dragon.[28] This applique, whose dimensions are approximately 8 x 5cm, has been found without any evidence of original inlays, probably removed before the remelting. The piece, in fact, was discovered, distorted by fire in antiquity, within a room near a hearth 35cm. in diameter, whose interior surface was hardened from intense heating. This fact suggests that the gold was destined for remelting and reuse of the main matter after the disassembling of all the components.

Polymateric work constitutes, in some cases, the capacity to merge reality and imagination. The activity of 'built' things that are not immediately perceived through the senses (as the aforementioned lion-demon plaque) derives from the need to realise objects known, above all, from myths and beliefs. Some artefacts discovered in the Royal Tombs at Ur, for instance, could be associated with images derived from myth, as a group of gold and lapis lazuli beads in the shape of flies strung together to form a necklace and maybe referring to the words pronunced by goddess Nintu in the Epic of Atrahasis.[29] Artefacts originated from the mental pictures described in written (and/or oral) narratives are experienced through memory, perception and imagination. On the one hand, memory is vital in the creative process. New ideas always rely on stored information for their 'creation'. The sensory stimulation provided by the memory of the human experience is also fundamental to the creative action: senses collaborate closely to enable the mind to better understand its surroundings. The imagination is

---

[22] Cifarelli 2010: fig. 2, note 11.
[23] See also some considerations by Suter 2000: 57; Ornan 2007: 448.
[24] On the archaeological context, see Omura 2008: 1-44.
[25] Omura 2015: 9-22.
[26] Boccia Paterakis, Omura and van Bork 2015: 106, figs. 1-2.
[27] Boccia Paterakis, Omura and van Bork 2015: 106-107. For a different opinion, see Omura 2015: 12.

[28] Boccia Paterakis, Omura and van Bork 2015: fig. 2.
[29] Hansen 1998: 47 and no. 69.

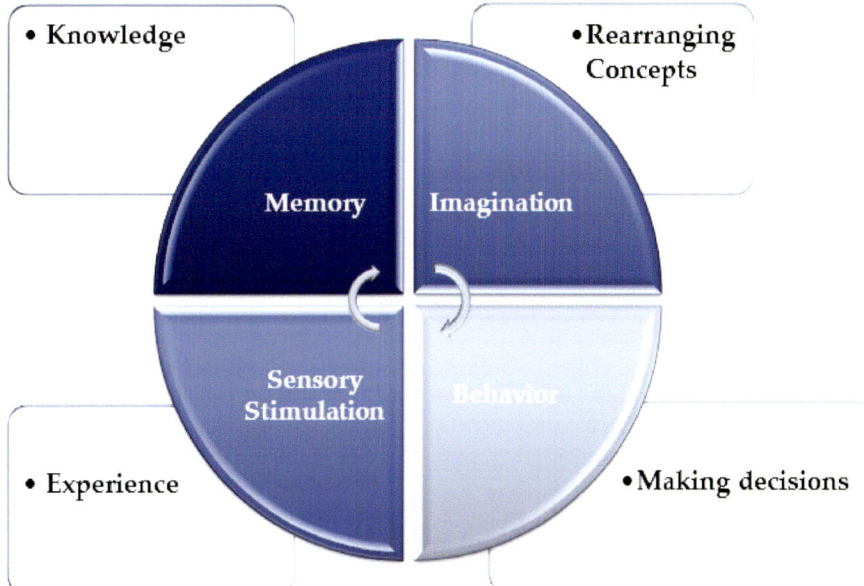

Figure 3. The creative process: materiality, memory and imagination (Treadaway 2009; final image by the author).

always 'structured' because the features of new ideas are arranged in predictable ways by the properties of existing categories and concepts: they do not arise in a vacuum.[30]

The visual representations and artefacts generated by them are 'inspired' by the way in which the world is experienced and perceived, remembered and imagined (Figure 3).[31]

The divine beings described in mythology commonly dwell in realms outside of the human experience since time immemorial: for example, in Mesopotamian mythology, the separation between human and divine was built into the fabric of creation and the Enuma Elish. Although the interpenetration between history and other forms of representation of the past (above all, mythology) is a constant element in the ancient Near East,[32] artisans are nevertheless constantly involved in the creation of things that not only should serve specific purposes (furniture, vessels, ornaments, etc.) but have to acquire particular divine attributes and embody vitality and truth in the eyes of the observers and users. In many cases, the main goal is to produce temple/palatine furniture or movable cultic furnishing. Sacred images, housed in temples, were sculpted in different materials, also assembled together (wood, metal, stone, etc.). They embodied divine, human, and mythological beings, as well as animals or specific attributes with their own divine nature or symbols of specific deities (divine weapons, for instance).[33]

On the floor of the Level VIIB of the Temple of Inanna at Nippur (Early Dynastic I-II) was found an important ceremonial item used for specific ceremonies in honour of the Mesopotamian goddess. It is a ritual vase in soapstone (a grey-green soft stone) characterised by the addition of small pieces of coloured stones and mother-of-pearl. Hundreds of similar vessels (incrusted, simply incised, or in relief) have been discovered throughout the ancient Near East, from Syria to the Indus Valley. They were locally produced in the southern part of Kerman (Iran), localised in the region of the Strait of Hormuz and the Persian Gulf. This particular production, known by specimens also found in southern Mesopotamia in 3rd millennium BC contexts, shows the existence of complex exchanges between this area and various regions of Iran.[34] The vessel from Nippur, now preserved in the Iraq Museum (IM 66071), is conical in form, tapering from its base to the neck and with a flaring wide-mouthed rim (Figure 4). The preserved piece is formed by the lower part, characterised by the animalistic decoration of a leopard and a snake. A feline in profile and facing left is grasping a snake. The leopard fur, in nature, is marked by black rosettes densely packed, while on the vase, it is depicted with numerous inlays of mother-of-pearl and orange/red stones set in circular holes on the body (these colours contrast with the darker surface of the background).[35] There is particular attention to the realistic rendering of the animal fur. Rectangular inlays are applied on the tip of its tail that, in nature, is only partially spotted, having parallel bands towards the end. The snake's muzzle faces the leopard, while

---

[30] Ward 1995, 157-178.
[31] On these processes, see also Treadaway 2009.
[32] On this aspect, see Di Paolo 2016b: 144.
[33] See Selz 1997: 167-209 and Porter 2009: 153-194.

[34] Perrot and Madjidzadeh 2004: 1105-1120; 2005: 123-152; 2006: 99-112.
[35] Hansen and Dales 1962: 79; Land 1985: 365, no. 55, fig. a p. 306.

Figure 4. Fragment of a soapstone vessel from the Temple of Inanna, Nippur (Level VIIB). Iraq Museum 66071 (Land 1985: 365, no. 55, fig. a p. 306).

its body curls around the whole field. The original lattice design is indicated with two lines of oval-shaped holes inlaid with red and green stones, and mother-of-pearl which reproduces the iridescence of the snake skin.

It is uncertain if these animals are, in this case, personifications of contrasting forces of the nature that try to dominate, perhaps identified with some divine principle by individuals and society produced similar artefacts and sculpted such figures. But the boundary between imagination and the real world is very subtle. If whole experience of life including knowledge, memory and belief is filtered through the mind, continually projecting a sense of meaning onto people and things, on the other hand the real world and its living creatures help to explore the nature of knowledge. How can we tell whether our own mental images filtered through knowledge are accurate or vivid if we have no direct comparison? That is, how do we come to know and judge the contents of our own minds? The common distinction between imagination and perception, with the former indicating something that 'is not there' and the latter signifying 'what is there' is, by itself, not helpful in determining the 'realness' of a mental experience[36] (Figure 5).

Panthera pardus ciscaucasica, commonly known as Persian leopard, is a subspecies native to Iran, and also to Northern Iraq.[37] Their presence in the Zagros area from the proto-historic period is demonstrated by numerous specimens of stone figurines discovered in southern Mesopotamia. A group of leopard figurines was found at Uruk: the flower-like pattern of black spots was indicated by pairs of crescent-shaped holes (they reproduce the different fur colours on the exterior borders of the spots) and then filled in with semi-precious stones.[38] Similar figurines from the Ubaid period have painted spots,[39] while a walking leopard has been discovered at Nippur.[40] Seals also in the shape of a leopard have been discovered at Tell Hamoukar: it is lying with his front paws outstretched, a very natural pose for this animal. Here, the spots are indicated by a series of drilled holes that were filled with another material.[41] The protective role of this animal has been suggested by to documentary sources and images: in particular, the scenes of the leopards fighting snakes indicate the positive role of the former, sometimes representing deities struggling with negative forces, protecting and enhancing the living conditions of men. In this framework, leopard skins, particularly 'traited' in the small sculptures before examined with the insertion of other materials assume specific meanings, including an apotropaic function in an identifying context to such an extent to be used to cover the human garments emphasising the status and role of warriors, dignitaries, etc.[42]

The category of polymateric artefacts assembled by **juxtaposition** is characterised by a balance of single elements/materials adopted. The various components are assembled to all appear in full view. In this instance, each component has a specific role: they do not simply appear without changing the basic structure of the object but, on the contrary, with their intrinsic qualities, compose the design and define its final shape. These artefacts are realised through a simultaneous mixture of heterogeneous materials, or from different qualities of the same material. The principal function and effect of this balanced combination seems polycromy but, studying the combination and provenance context of each artefact, it is probable, in some cases, that the simultaneous mixture of different materials is structural (matters are used to compose the same figure), depending on function and meaning of the object, i.e. whether it was specifically made for

---

[36] O'Connor and Aardema 2015: 233-256.
[37] Kiabi, Dareshouri, Ali Ghaemi and Jahanshahi 2002: 41-47.
[38] Wrede 2003: 61 and figs. 27e-g. The insertion is also used on clay figurines: Wrede 2003: 219 and figs. 838-839.
[39] Heinrich and Hilzheimer 1936: pl. 13F.
[40] Legrain 1930: 35 no. 320.
[41] Nys and Bretschneider 2007: 17, no. 44.
[42] Nys and Bretschneider 2007: 555-615.

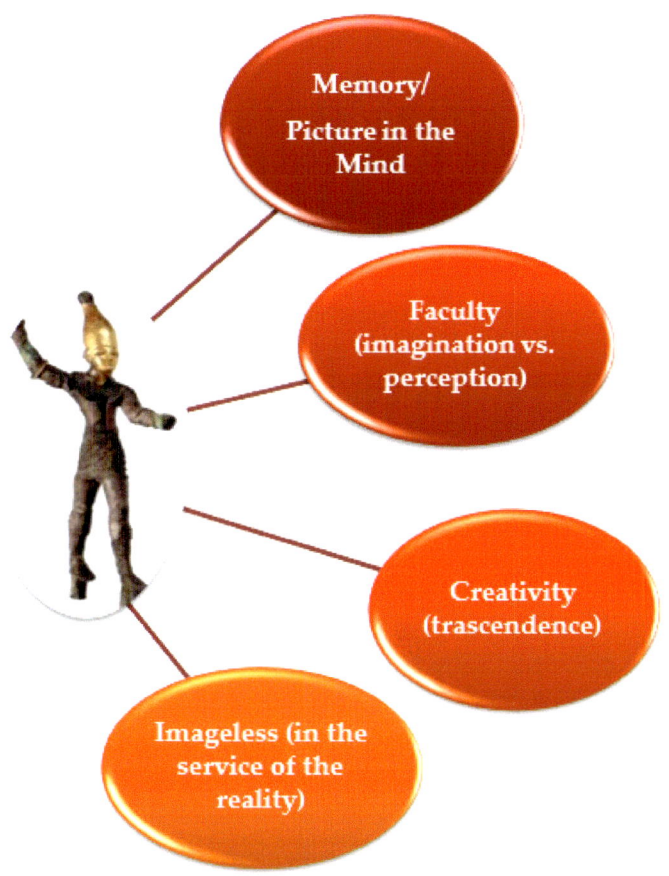

Figure 5. Different approaches to the concept of imagination (O'Connor and Aardema 2005: 234, Table 1; final image by the author).

much rarer in neighbouring areas (Egypt, for instance), where they only appear in a restricted range of designs. The final Middle Bronze Age marks the beginning of a long tradition of bone carving in Palestine (Tell Beit Mirsim, Lachish, Tell el-Fa'rah (S), Hazor, Jericho, Gezer, Tel Dan).[45] Most inlays are rectangular, but there are also silhouette forms, cut out in the shape of human figures, various types of animal (birds, felines, and mammals, for instance) and symbols (hieroglyphic signs) and geometric: concentric circles, guilloche, chevrons, zig-zag lines, herringbone pattern, and other incised line designs are common treatments on rectangular and sometimes triangular strips. From a tomb at el Jisr comes a collection of ivory fragments which were part of a complex inlay. Although too fragmentary, the composition contained human figures, and animal such as lions, horses, oxen, and birds. The strips, used to frame panels and designs, were pierced for attachment by pegs, while the silhouettes were, generally, glued on the background.[46]

religious and funerary purposes. Bone and ivory inlays applied on a wooden support to form furniture or small containers constitute one of the most common categories of artefacts realised by juxtaposition. They are characterised by the application of figurative or geometrical elements to a structure to form decorative patterns or designs: the prevalent aim is the contrast of materials and colours for ornamental purposes. This technique was applied to furniture or small objects with a smooth surface, or again to freestanding pictorial panels. It differs from the insertion, in which a solid body of one material is cut out to receive sections of another to form the surface pattern.

One of the more common finds in settlement and burial contexts of Syria and Palestine in the final Middle – beginning of the Late Bronze Age are bone inlays featuring a range of designs incised and, sometimes, inlaid with black pigment to stand out the colour of the bone and, probably, the wooden background. Ivory inlays are common in Syria and Anatolia in this period (Ebla, Alalakh, Kültepe-Kanesh);[43] bone versions are characteristics of Palestinian assemblages,[44] but are

In some cases, categories 3 and 4 (see after) overlap. Excavations by L. Woolley at Alalakh produced hundreds of geometric shapes and fragments for inlaying furniture, still stored in the dig-house depot located on the mound and described by K.A. Yener.[47] Some of these bone pieces were originally scored in order to permit the adherence of a thin silver foil caught in the incised grooves, discovered through SEM analyses.[48] Thus, the assemblage of different colours and materials (black/brown wood, white/beige bone and gray/white silver) showed, through recent diagnostic tools, that the visual sensation associated with silver looked for by users is its metallic shine. As with gold, this cannot be reproduced by a simple colour; the shiny effect is due to the material's brightness varying with the surface angle to the light source of decorative panels and furniture inlaid.

A particular group of artefacts is characterised by mixing all of the aforementioned techniques. In some cases, in fact, polymateric objects are conceived by combining a high number of techniques and materials,

---

[43] Scandone Matthiae 2002; 2006; Yener 2007: especially 153-156; Özgüç 1986: 71, pls. 123-124, fig. 52.
[44] Leibowitz 1977: 89-97.
[45] Albright 1938: 56-58; Kenyon 1960; Tufnell 1958: 86-87, pl. 28.1-4; Pritchard 1963.
[46] Amiran 1977: 65-69.
[47] Yener 2002: 13-19.
[48] Yener 2007: 153.

partly visible, partly hidden. The polymateric quality is, probably, not only intended for aesthetic ends, but also tied to the transmission of symbolism of these special artefacts. Its implications must have arrived with force and clarity by taking such extraordinary forms, characterised, in a certain sense, by 'divine' mixture which induces a 'profound' reflection. The game of using and mixing materials and colours according to expressive semantic meanings and values which they produced extended beyond this. They produce deep sources of symbolism which significantly enhanced their meaning and functions. In these cases, the investigation about choice and use of materials has to be evaluated while considering the strong symbolic congruity of the material from which the artefact is made and its color as well as its precious character. Every material and every color has a semantic meaning, creating an iconology of material and colour.

In the ancient Near East, ivory production fully meets the characteristics aforementioned. As is known, the Levantine palatial culture of the Early Iron Age is represented, *inter alia*, by the production and consumption of ivory objects. Other relevant specimens dating back to the Late Bronze Age are an index of the continuity of tradition and taste within power elites, although developments and inter-regional connections of these different productions remain unclear. Composite objects in ivory, made using multiple materials and multiple techniques, would have created a form of distributed 'cognition' involving the coordination between individuals, artefacts and the environment.

Firstly, materials and their properties are carefully explored by craftsmen for the purpose of planning manufacture: this exploration spans from the availability and choice of materials, to assembling techniques, and blending contrasting properties, qualities, and colours. Brought together in accumulation, these artefacts become a meaningful aggregate of associations and spatio-temporal relationships.[49] Composite construction appear to crosscut regional and stylistic boundaries: it might be regarded as the fundamental tenet of an overarching symbolic system based on different materials, colours and their capacity to define form and create spatial illusion.

Multi-component ivories, which are characterised by the use of different materials (ivory, stone, gold, glass) and colours, are common artefacts of the Iron Age. They have been found particularly at Nimrud, Arslan Tash, and Samaria. At Nimrud, composite construction is seen in a series of openwork panels attributed to the Ornate Group belonging to the Classic Phoenician tradition, but it is also known at Arslan Tash and Samaria.[50] The Egyptianising style of all these ivories, and the common use of coloured inlays, has led to the hypothesis that only Phoenician craftsmen were experts at inlaying other materials into ivory (gold and glass pieces) with the aim of creating composite constructions, and that each processing stage was developed and finished by them.[51]

However, evidence contradicts this. An oval pyxis discovered in the Well AJ of the North-West Palace at Nimrud, belonging to the Flame and Frond School (one of the North-Syrian stylistic groups), was carved in high relief with two similar scenes. Each shows a nude winged mistress of animals (probably a divine figure) grasping a hind leg of each of a pair of lions.[52] The female figure is represented frontally, her wings outspread and formed by two rows of cloisons which were originally inlaid. The inlays have almost completely disappeared; the few examples which are still preserved *in situ* show another way of interpreting the concept of the composite artefact. The divine wings end in volutes originally inlaid with 'brown' ivory inlays (blackened or burned on purpose?). This pyxis shows the adoption of two completely new solutions in comparison to the Phoenician specimens: the use of ivory as an inlay material (probably from wasted pieces) and the darkening of the inserts in order to create a *chiaroscuro*, without changing the nature of the material.

In other cases, the large cloisons, forming the four wings of the mistress of animals, were filled with pieces of brown coloured inlay (traces of brown paste are *in situ*) and burnt ivory.[53] It is uncertain if the gold foil overlaid, in some parts, the area of the cloisons. The state of conservation of these artefacts is, often, mediocre. Nevertheless this aspect seems relevant: if the gold leaf was hammered over the cloisons filled with inlays, it is possible to explain the use, in some cases, of simple ivory inserts instead of multi-coloured pastes made in different materials to be seen. The use of gold and coloured glass or stone pieces was probably aimed at strengthening the symbolic potential of ivory. When used in abundance for the walls of cloisons, glittering and shining gold extends its qualities to the entire figure,[54] while stone and glass pieces amplified the composite nature of the artefact by assembling hair, garment, and body parts in a 'imaginary' world.[55] Through the improved materiality of these artefacts,

---

[49] These aspects have been studied by author in two articles: Di Paolo 2016b; Di Paolo in press.

[50] Herrmann 1986: 20–22; 1992, 35–37; Herrmann and Laidlaw 2013: 26–56; Affanni 2012: 193–208; Fontan and Reiche 2011: 283–295; Suter 2011: figs. 1–2.
[51] Barag 1993: 1–8.
[52] Safar and al-Iraqi 1987: 59, no. 11; Herrmann and Laidlaw 2009: 186–187, no. 235, pls. 48–49.
[53] Safar and al-Iraqi 1987: 48–53, no. 9; Herrmann and Laidlaw 2009: 185, no. 234, pls. 44–47.
[54] Herrmann and Laidlaw 2009: 165–166, pls. F–G, 23 no. 158.
[55] Herrmann 1992: pl. LXVIII, no. 496.

including the use of gold and stone or glass (substitutes for genuine stones), objects are saturated with meaning. In the interaction between humans and artefacts, the art of crafting is a virtuosity that exemplifies an ideal or magical efficacy.

At the Intersection between Materiality and Knowledge: the Body Metaphor

Composite artefacts can be considered the result of a 'special' interaction. On the one hand, the material sphere (the external world) is where materials exist in nature and are modified by humans, are adapted, transformed, and assembled to produce a finished object. On the other hand, there is the cognitive sphere in which the significant evidence of planning objects made of parts from dispersed sources effects a profound change in the way information is processed. Over the last decades, research has highlighted how composite tools represent, since the Neanderthals, a creative association between structures of the brain involved in performing hierarchically constructive actions: attention, memory, and imagination.[56] It is not only a matter of tool-construction process requiring a longer sequence of motor actions or a multistage assemblage process (also extending over days/weeks).[57] This kind of artefacts makes special demands on the cognitive system because they also have a greater potential to accumulate meanings and connections than single objects because of their 'assembled biographies'. It has also been suggested that the concepts of culture, brain, body, and mind can only exist as an ontological unity, implying that culture envelops and constitutes human cognitive and emotional lives. Whether or not, as Lambros Malafouris explains, the material world plays a very important role in the structuring of human cognitive operations, nevertheless we are still a long way from arriving at an 'empirically testable formulation' of the key issues in this specific field of enquiry.[58]

An important aspect about the boundary between perception and imagination is that both can exist simultaneously, and this rather upsets the notion that one concerns 'non-reality' and the other 'external reality'. People can easily imagine without losing contact with reality, occupying both imaginary and real space at the same time: people can function whilst not completely in reality nor in the imagination but somewhere in between.[59] In particular, mixing ideas, beliefs, materials, techniques, and assembling methods, polymateric artefacts appear as the result of a mediation between: 1) the capacity to create 'things' from mental pictures originating in individuals' memory and knowledge and 'experienced' through imagination (the activity of imagining things that are not perceived through senses); and 2) the ability to efficiently and accurately assemble a combination of different components. I think that, in many cases, skill works together with sets of beliefs, stories and myths embodied in a large collection of narratives which explain the origins of the world, the characteristics of the divine bodies and other mythological creatures, and the symbolism and ideology of royal power. Through these processes, finished artefacts gain vitality, as well as potentially embodying of a living truth in the eyes of the viewer.

One of the best examples illustrating these concepts is the representation of Imdugud/Anzû during Early Dynastic and Akkadian periods. It is visually associated to a composite figure of lion-headed eagle, known in several exemplars in Mesopotamia and Syria. At Mari, within the foundation deposit known as the 'treasure of Ur' in the sanctuary associated to the palais P 0 (third quarter of the 3rd millennium BC) was discovered the composite figure of a *lion*-headed eagle: body and wings are fabricated in lapis lazuli, heads and tails in gold.[60] (Figure 6). The divine body assemble precious metals and stone together. The symbolism here seems prevalent: lapis lazuli and wings could refer to 'unspecific' sky, although the divine bird is specifically associated to the dark clouds rolling across the sky, as recalled in the written sources.[61] The head and tail seem strictly linked to the identity/nature of the divine being (a lion-headed bird), rendered with a metal shining foil. In particular, the narrow tail could indicate the speed during flight. Normally, as a bird accelerates it gradually reduces its tail's angle in relation to airflow, and at higher speeds it furls its tail.[62] Thus, this representation could have promoted the image of a bird that quickly cross the sky like the rain clouds, rather than displaying the roaring creature, another aspect highlighted by texts.

As regards materials and their natural properties, including colours (modified and enhanced through manufacturing processes) and shapes, they are used for promoting and extolling virtues and qualities. Artisans amplify the composite nature of the artefact by assembling garments, body parts, and landscape in a uniquely imaginary world: contrasts of colours and the interplay of transparency or opacity illuminate the materials around them, and exemplify the value of polychromy. However, 'polychrome' is a separate colour concept: in the ancient Mesopotamian texts, this term focuses on the ideas of 'variegation' and 'patterning' of specific materials (for instance, it is applied to

---

[56] About this, see the relevant remarks made by Wragg Sykes 2015: 117-137 (with preceding bibliography).
[57] Rossano 2010.
[58] Malafouris 2010: 264.
[59] O'Connor and Aardema 2015: 233-256.
[60] Margueron 2004: 537, pl. 70.
[61] On this aspect, see the article by Ch. Watanabe in this volume.
[62] Biewener 2011: 1496-1506.

Figure 6. Composite lion-headed aigle from Mari. 'Treasure of Ur'. Palace P-0. Third quarter of 3rd millennium BC. (Margueron 2004: 537, pl. 70).

embroidered textiles), and brings together all of the meanings of the colours, emphasising the congruity of parts to their whole, as well as a higher value for the multiplicity.[63] Composite artefacts represent appointed places at which to build heterogeneous relationships between the most disparate and incongruous elements. They have more concrete relationships with the external world; however, they emphasise at the same time 'the creative faculty of the mind' as being the synthesis between thought and images, between 'externalised' reality and 'internalised' man.

In this network of associations, the body and its construction represents a powerful metaphor, perhaps *the* most powerful metaphor. The body is simultaneously a physical and symbolic artefact, being both naturally and culturally produced, and is securely anchored in a specific historical moment. At the first and perhaps most self-evident level is the individual body, understood in the phenomenological sense of the lived experience of the body-self. At the second level, the social body refers to the representational uses of the body as a natural symbol with which to think about society and culture: there is a constant exchange of meanings between the natural and social worlds. The body offers a model of organic wholeness, while in particular situations (wars, for instance) it is associated with the concept of disharmony and/or disarticulation.[64]

Iconic representations of human/animal bodies are, in some cases, conceived as the result of a combination of isolable but interactive parts. The intangible cohesion of the body is replaced by a divisible body made up of mutually involved but separate parts. The metaphor of the natural/social body allows us to build it as a composite artefact, formed of several parts reproducing an integrated whole. Thus, it becomes a kind of 'fabric', at first mentally disarticulated and then reassembled as a multi-component object. But key points are: which 'body part' is to be realised in another material? How are single parts selected? What do specific contexts reveal about cultural (de)constructions of bodies? How is the body fragmented or deconstructed? What do such (di)sarticulations indicate about the integrity of the body and the social worth of divine, human or animal representations?

Terracotta hands, in some cases glazed, have been discovered from the collapsed walls of the Nimrud palace. Flat on one side, they are moulded; on the opposite side, and represent closed fists but with clearly visible fingers.[65] Other specimens of hands obtained from bronze foils hammered on a mould or simply incised with pointed tools come from the palaces of Nimrud and Nineveh.[66] Provided with cuneiform inscriptions inscribed on them with the name and titles of the king responsible of their manufacture, these body fragments, perhaps hammered to the wood or hollow inside, were hung on the walls probably with a ritual function.[67] Faces distinguished in material to the rest of the body are produced in order to build a 'variegated' colour palette exalting the light skin tone against the darker colour of hair and eyebrows, as well as the importance of eyes inserted in a different material: a general effect that we cannot never appreciate. A stone mask discovered in the area of the Massif Rouge at Mari, probably covering a wood core, shows areas for incrustations of hair, eyes and eyebrows. In particular, the eye contact with the viewer is a powerful and penetrating form of communication that radiates all over the face of the real and metaphorical body also thanks to the contrasting materials and colours, as well as the fascination for technology.

The process of disarticulation seems strictly linked to the symbolic construction (reconstruction) of the body, which may inform viewers how the world is thought and imagined, and how and why such mental (de)constructions and practical reconstructions augment or alter the social worth of the body represented. To summarise, body fragments such as hair, eyes, faces, hands etc. may be emotionally charged objects embodying transformative properties that have the power to convey emotions, fear, admiration, and so

---

[63] Warburton 2007, 229–246; Sinclair 2012, 9–11.
[64] For a specific aspect, see Di Paolo: in press.
[65] Preusser 1955: pl. 14b.
[66] Curtis 2013: 63–64.
[67] Frame 1991: 335–381; Neumann 2014: 307.

Figure 7. Detail of a gaming board from Ur. Tomb PG 580. Early Dynastic IIIA. 2550-2400 BCE. (Hansen 1998, 60, fig. 6 below).

on. The eyes, for instance, are used for a multiplicity of aims, including as 'conceptual' and material borders.[68] Bodies generate a host of potent metaphorical constructions for ordering the world. Materials and the visual language employed to 'describe' single parts of the body reflect how they are understood in ethical and social terms.

Inventories of the furniture in the temples or shrines include, in many cases, descriptions of the god's image. Hittite cult inventories preserved in the archives at Hattusa describe the body of the divine statue as well as associated materials in some detail. For example, the image of the goddess Iyaya worshipped in the Anatolian town of Lapana (KUB 38.1 iv 1-7) is indicated by a wooden statue (40cm. c high) of a woman seated and veiled. Her head is inlaid with gold, while her body and throne are covered with tin. Two wooden sheep inlaid with tin are seated beneath the goddess, to the right and left.[69]

The construction of the statues generally intended in the ancient Near East as imbued with divine essence (in this case the term is *šiuniyatar*, an abstract of the Hittite word for deity = *šiuni-*), is represented as a body in 'parts', that is a body constituted by a multiplicity of individual sections. In the description of a gold statue of the king Hattusili to be devoted to the goddess Lelwani, head, hands and feet are clearly distinguished.[70]

This 'fragmentation', strictly linked to the description of the materials employed, lead to the accentuation of the individual character and importance of each part. Thus, the single parts are not pieces of a shattered whole, but individual units of the body in which specific aspects of culture are imagined to reside.

Although it is impossible to exactly reconstruct the aspect of the wooden statue of Iyaya, from the description it can be ascertained that the head is clearly distinguished from the rest of the body. The head is

---

[68] On the value of the eyes in the polymateric work at Qatna, see the article by E. Roßberger in this volume.
[69] Collins 2005: 16.

[70] Goetze 1969: 394.

overlaid with gold (was the veil perhaps formed by a tissue covering it?), while the rest of the body (torso and legs?) is intended, as a 'compact unity', overlaid with tin, a highly crystalline silvery-white metal and a less expensive material compared to gold. The location of the statue inside the temple with respect to the sources of light and its angle, would have emphasised the reflectivity of the metals, particularly making tin a mirror surface also amplified by the sheeps covered with tin located on both sides of the divine throne. Furthermore, the conception of the human body as a whole of integrated parts and therefore each possessing a specific 'meaning' of their own is associated with the sitting position of the goddess that emphasizes stability (no hint of movement) aside from focusing on the body-throne binomy. All of the examples mentioned here showed that the composite body is never a unified entity, but always a 'fragile' picture resulting from the piecing together of dissimilar parts. Many key issues remain on the table. This article has only tried to raise questions and doubts exploring different categories of composite artefacts from different regions of the ancient Near East. This topic is intriguing but also problematic and complicated. But it is worth to investigate it further in the next future.

References

Affanni, G. 2012. New Light (and Colour) on the Arslan Tash Ivories: Studying 1st Millennium BC Ivories. In R. Matthews and J. Curtis (eds), *Proceedings of the 7th International Congress on the Archaeology of the Ancient Near East, 12-16 April 2010*: 193-208. Wiesbaden: Harrassowitz.

Albright, W.F. 1938. *The Excavations at Tell Beit Mirsim, vol II: the Bronze Age*, AASOR, vol. 17. New Haven (CT): American Schools of Oriental Research.

Amiran, R. 1977. The Ivory Inlays from the Tomb at el-Jisr Reconsidered, *Israel Museum News* 12: 65-69.

Barag, D. 1993. Glass Inlays in Phoenician Ivories, Glass and Stone Vessels, *Annales du 12e Congrès AIHV*: 1-8.

Biewener, A.A. 2011. Muscle Function in Avian Flight: Achieving Power and Control, *Philosophical Transactions of the Royal Society B: Biological Sciences* 2011/366: 1496–1506.

Boccia Paterakis, A., Omura, S. and E. van Bork 2015. An Unusual Example of Gold Cloisonné from Central Anatolia, *STAR: Science & Technology of Archaeological Research* 1/2: 106-114.

Cifarelli, M. 2010. Adornment, Identity, and Authenticity: Ancient Jewelry in and out of Context. Exhibition Review, *American Journal of Archaeological Online Museum Review* 114: 1-9.

Cluzan, S. 2017. Nouvelles observations iconographiques et techniques sur la statuaire du Bronze d'Ugarit: premiers résultats du project d'étude et de restauration mené en partenariat entre le Musée du Louvre et le Musée National de Damas. In V. Matoïan (ed.), *Archéologie, patrimoine et archives. Les fouilles anciennes à Ras Shamra et à Minet el-Beida I*, Ras Shamra-Ougarit 25: 71-89. Leuven, Paris and Bristol: Peeters.

Collins, B.J. 2005. A Statue for the Deity: Cult Images in Hittite Anatolia. In N.H. Walls (ed.), *Cult Image and Divine Representation in the Ancient Near East*: 13-42. Boston: American Schools of Oriental Research.

Curtis, J. 2013. *An Examination of Late Assyrian Metalwork with Special Reference to Nimrud*. Oxford: Oxbow Books.

Di Paolo, S. 2016a. War Remembrance Narrative: Negotiation of Memory and Oblivion in Mesopotamian Art. In D. Nadali (ed.), *Envisioning the Past through Memories. How Memory Shaped Ancient Near Eastern Societies*: 143-162. London: Bloomsbury.

Di Paolo, S. 2016b. Beyond Design and Style: Enhancing the Material Dimension of Artefacts through Technological Complexity. In C. Suter (ed.), *Levantine Ivories of the Iron Age: New Perspectives. 61th Rencontre Assyriologique Internationale = Text and image*, Geneva and Bern, 22-26 June 2016, AoF 42/1: 71-79.

Di Paolo, S. in press. Perception and Appreciation of the Materiality: Levantine Multi-component Ivories. In M.G. Micale et al. (eds) *A Oriente del Delta. Scritti sull'Egitto e il Vicino Oriente antico in onore di Gabriella Scandone Matthiae*. Rome: 'Sapienza' University of Rome.

Douny, L. and S. Harris 2014, Wrapping and Unwrapping. Concepts and Approaches. In S. Harris and L. Douny (eds) *Wrapping and Unwrapping Material Culture. Archaeological and Anthropological Perspectives*: 15-40. Walnut Creek: Left Coast Press.

Fontan, E. and I. Reiche 2011. Les ivoires d'Arslan Tash (Syrie) d'après une étude de la collection du Musée du Louvre: mise en oeuvre du matériau, traces de polychromie et de dorure, état de conservation, *ArchéoSciences* 35: 283–295.

Frame, G. 1991. Assyrian Clay Hands, *Baghdader Mitteilungen* 22: 335-381.

Goetze, A. 1969. Hittite Prayers. In J.B. Pritchard (ed.), *Ancient Near Eastern Texts Relating to the Old Testament*: 394-396. Princeton: University Press.

Güterbock, H.G. 1957. Narration in Anatolian, Syrian, and Assyrian Art, *AJA* 61: 62-71.

Hansen, D.P. and G.F. Dales 1962, The Temple of Inanna, Queen of Heaven at Nippur, *Archaeology* 15: 75-84.

Hansen, D.P. 1998, Art of the Royal Tombs of Ur: A Brief Interpretation. In R.L. Zettler and L. Horne (eds), *Treasures from the Royal Tombs of Ur*: 43-72. Philadelphia: University of Pennsylvania Museum of Archaeology and Anthropology.

Heinrich, E. and M. Hilzheimer 1936. Kleinfunde aus den archaischen Tempelschichten in Uruk, *Ausgrabungen der deutschen Forschungsgemeinschaft in Uruk – Warka I*. Leipzig: Harrassowitz.

Herrmann, G. 1986. *Ivories from Nimrud (1949-1963) IV. Ivories from Room SW37, Fort Shalmaneser*. London: The British School of Archaeology in Iraq.

Herrmann, G. 1992. *Ivories from Nimrud (1949-1963) V. The Small Collections from Fort Shalmaneser*. London: The British School of Archaeology in Iraq.

Herrmann, G. and S. Laidlaw 2009. *Ivories from Nimrud (1949-1963) VI. Ivories from the North West Palace (1845-1992)*. London: The British School of Archaeology in Iraq.

Herrmann, G. and S. Laidlaw 2013. *Ivories from Nimrud (1949-1963) VII. Ivories from Rooms SW11/12 and T10 Fort Shalmaneser*. London: The British School of Archaeology in Iraq.

Kenyon, K.M. 1960. *Excavations at Jericho II. The Tombs Excavated in 1955-1956*. Jerusalem: British School of Archaeology in *Jerusalem*.

Kiabi, B.H., B.F. Dareshouri, R. Ali Ghaemi, and M. Jahanshahi 2002. Population Status of the Persian Leopard (Panthera pardus saxicolor Pocock, 1927) in Iran, *Zoology in the Middle East* 26: 41–47.

Land 1985. *The Land between Two Rivers. Twenty Years of Italian Archaeology in the Middle East. The Treasures of Mesopotamia*. Turin: Il Quadrante.

Lechtmann, H. 1971. Ancient Methods of Gilding Silver: Examples from the Old and the New Worlds. In R.H. Brill (ed.), *Science and Archaeology*: 2-31. Cambridge (Mass.): MIT Press.

Legrain, L. 1930. *Terra-cottas from Nippur*. Publications of the Babylonian Section XVI. Philadelphia: University of Pennsylvania Press.

Liebowitz, H. 1977. Bone and Ivory inlay from Syria and Palestine, *Israel Exploration Journal* 27: 89–97.

Loud, G. 1936. *Khorsabad I. Excavations in the Palace and at A City Gate*. OIP 38. Chicago: University Press.

Luckenbill, D.D. 1924. *The Annals of Sennacherib*. Oriental Institute Publications 2. Chicago: University of Chicago Press.

Luckenbill, D.D. 1926-7. *Ancient Records of Assyria and Babylonia*. Chicago: University Press.

Malafouris, L. 2010. The Brain–Artefact Interface (BAI): a Challenge for Archaeology and Cultural Neuroscience, *Social Cognitive and Affective Neuroscience* 5: 264–273.

Margueron, J.-C. 2004. *Mari. Capitale sur l'Euphrate*. Paris: Picard.

Meller, H., R. Risch and E. Pernicka (eds) 2014. *Metalle der Macht – Frühes Gold und Silber. Metals of ower – Early gold and silver. 6. Mitteldeutscher Archäologentag vom 17. bis 19. Oktober 2013 in Halle (Saale)*. Tagungen des Landesmuseums für Vorgeschichte Band 11/1: 11-17. Halle (Saale): Landesamt f. Denkmalpflege u. Archäologie Sachsen-Anhalt.

Neumann, K. 2014. *Resurrected and Reevaluated: the Neo-Assyrian Temple as a Ritualized and Ritualizing Built Environment*. PhD Diss., University of California: Berkeley.

Nys, N. and J. Bretschneider 2007. Research on the Iconography of the Leopard, *Ugarit Forschungen* 39: 555-615.

O'Connor, K.O. and F. Aardema 2005. The Imagination: Cognitive, Pre-cognitive, and Meta-cognitive Aspects, *Consciousness and Cognition* 14: 233–256.

Omura, S. 2008. Preliminary Report on the 22nd Excavation Season at Kaman-Kalehöyük in 2007, *Anatolian Archaeological Studies* 17: 1–43.

Omura, M 2015. A Gold Plaque from Kaman-Kalehöyük and the 'Lion-Dragon' Motif, *Anatolica* 41: 9–22.

Ornan, T. 2007. The Role of Gold in Royal Representation: the Case of a Bronze Statue from Hazor. In R. Matthews and J. Curtis (eds), *Proceedings of the 7th International Congress on the Archaeology of the Ancient Near East, 12 April – 16 April 2010, British Museum and UCL, London*, vol. 2: 445-458. Wiesbaden: Harrassowitz.

Özgüç, T. 1986. *Kultepe-Kanis II. New Researches at the Trading Center of the Ancient Near East*. Ankara: Türk Tarih Kurumu Basimevi.

Perrot, J. and Y. Madjidzadeh 2004. Récentes découvertes à Jiroft (Iran): résultats de la campagne de fouilles, 2004, *Comptes-rendus de l'Academie des inscriptiones et belles-lettres*: 1105-1120.

Perrot, J. and Y. Madjidzadeh 2005. L'iconographie des vases et objets en chlorite de Jiroft (Iran), *Paléorient* 31/2: 123-152.

Perrot, J. and Y. Madjidzadeh 2006. À travers l'ornamentation des vases et objets en chlorite de Jiroft, *Paléorient* 32/1: 99-112.

Policromia 2004. *I colori del bianco: policromia della scultura antica*. Rome: De Luca Editori d'Arte.

Ponting, M.J. 2013. Scientific Analysis of the Late Assyrian Period Copper-Alloy Metalwork from Nimrud. In J. Curtis, *An Examination of Late Assyrian Metalwork with Special Reference to Nimrud*: 220-242. Oxford: Oxbow Books.

Porter, B. 2009. Blessing from a Crown, Offering to a Drum. Were there Non-Anthropomorphic Gods in Ancient Mesopotamia? In B. Porter (ed.), *What is a God? Anthropomorphic and Non-Anthropomorphic Aspects of Deity in Ancient Mesopotamia*: 153-194. Winona Lake: The Casco Bay Assyriological Institute and Eisenbrauns.

Preusser, C. 1955. *Die Paläste in Assur*. Wissenschaftliche Veröffentlichung der Deutschen Orient-Gesellschaft 66. Berlin: Mann.

Pritchard, J.B. 1963. *The Bronze Age Cemetery at Gibeon*. Philadelphia: University Museum.

Renfrew, C. 1986. Varna and the Emergence of Wealth in Prehistoric Europe. In A. Appadurai (ed.), *The Social Life of things: Commodities in Cultural Perspective*: 141-168. Cambridge: Cambridge University Press.

Rossano, M.J. 2010. Making Friends, Making Tools, and Making Symbols. *Current Anthropology* 51/S1, S89–S98.

Safar, F. and M.S. al Iraqi 1987. *Ivories from Nimrud*. Baghdad: Republic of Iraq.

Scandone Matthiae, G. 2002. *Gli avori egittizzanti dal Palazzo Settentrionale*. Materiali e Studi Archeologici di Ebla. Roma: 'Sapienza' University of Rome.

Scandone Matthiae, G. 2006. Nuovi frammenti di avori egittizzanti da Ebla. In E. Czerny, I. Hein, H. Hunger, D. Melman and A. Schwab (eds), *Timelines. Studies in honour of Manfred Bietak* II: 81-86. Leuven, Paris, Dudley (MA): Peeters.

Selz, G. 1997. The Holy Drum, the Spear, and the Harp. Towards an Understanding of the Problems of Deification in Third Millennium Mesopotamia. In I. Finkel and M. Geller (eds), *Sumerian Gods and their Representations*: 167-209. Groningen: Styx Publications.

Sinclair, A. 2012. Colour Symbolism in Ancient Mesopotamia, *Ancient Planet Online Journal*, 3-15.

Suter, C.E. 2000. *Gudea's Temple Buildings. The Representations of an Early Mesopotamian Ruler in Text and Image*. Cuneiform Monographs 17. Groningen: Styx Publications.

Suter, C.E. 2011. Images, Tradition, and Meaning: the Samaria and other Levantine Ivories of the Iron Age. In G. Frame *et al*. (eds), *A Common Cultural Heritage: Studies on Mesopotamia and the Biblical World in Honor of Barry L. Eichler*: 219-241. Bethesda: CDL Press.

Treadaway, C. 2009. Materiality, Memory and Imagination: Using Empathy to Research Creativity, *Leonardo* 42/3, https://www.researchgate.net/publication/228683871_Materiality_Memory_and_Imagination_Using_Empathy_to_Research_Creativity [accessed Feb 15 2018].

Tufnell, O. 1958. *Lachish IV. The Bronze Age*. London: Oxford University Press.

Warburton, D.A. 2007. Basic Color Terms Evolution in the Light of Ancient Evidence from the Near East. In R.E. MacLaury, G.V. Paramei and D. Dedrick (eds), *Anthropology of Color: Interdisciplinary Multilevel Modeling*: 229-246. Amsterdam and Philadelphia: John Benjamins.

Ward, T.B. 1995. What's Old about New Ideas? In S.M. Smith, T.B. Ward and R.A. Finke (eds), *The Creative Cogniton Approach*: 157-178. Cambridge (MA): The MIT Press.

Winter, I.J. 2012. Gold! Divine Light and Lustre in Ancient Mesopotamia. In R. Matthews and J. Curtis (eds) 2012. *Proceedings of the 7th International Congress on the Archaeology of the Ancient Near East, London, 12 April - 16 April 2010, vol. 2*: 153-171. Wiesbaden: Harrassowitz.

Wragg Sykes, R. 2015. To See a World in a Hafted Tool: Birch Pitch Composite Technology, Cognition and Memory in Neanderthals. In F. Coward *et al*. (eds), *Settlement, Society and Cognition in Human Evolution. Landscapes in the Mind*: 117-137. Cambridge: Cambridge University Press.

Wrede, N. 2003. Uruk. Terrakotten I. Von der 'Ubaid-bis zur altbabylonischen Zeit. Mainz: von Zabern.

Yener, K.A. 2002. Tell Atchana. *The Oriental Institute 2001-2002 Annual Report*, 13-19.

Yener, K.A. 2007. The Anatolian Middle Bronze Age Kingdoms and Alalakh: Mukish, Kanesh and Trade, Anatolian Studies 57: 151-160.

A Composite Look at the Composite Wall Decorations in the Early History of Mesopotamia

Alessandro Di Ludovico
'Sapienza' University of Rome
alediludo@gmail.com

Abstract

The many complex processes that characterised the Urban Revolution have been variously outlined, but the mental and cultural steps that were combined with the relevant phenomena are rarely discussed, and above all elusive is their connection with the archaeological evidence. On the other hand, some types of wall decoration techniques that are attested in Lower Mesopotamia in the early historic period can play an important role in revealing the occurrence of remarkable cultural transformation processes. Paying a special attention to the logics and technical organization of some findings, like the Uruk wall cone mosaics, features can be located that serve as clues to explain the development of certain types of mental paths which have a quite direct relation with the dawn of the historical ages. The central effort of this contribution is dedicated to accomplish these tasks through methodological tools that are provided not just by neurosciences, but also (and more meaningfully) by old and new reflections in the fields of cultural studies.

Keywords: wall cone mosaics; Early Historic period; Uruk; visual perception; mental frames; cultural changes in Mesopotamia

As the first investigation activities at the site of Warka took place early on, a huge quantity of impressive finds were recovered. These gave researchers the opportunity to begin to outline and understand entire historical phases of ancient Mesopotamian cultures. Some of those finds can be included in categories that seem to be expressions specific to the late prehistoric or early historic peoples dwelling in this region. Well known and clear examples of these are the cretulae with cylinder seal impressions, and the earliest clay tablets with numerical signs or pictograms. The approximately contemporary polychrome wall mosaics made of cones are a type of artefact which should be connected to them. According to the underpinning logic, and the cultural inner workings which can be outlined for this historical age, this relationship is very strong, as will be shown here, and can illuminate interesting paths of research.

The Wall Cone Mosaics

During his first explorations at Warka in the mid-19th century, William K. Loftus reports finding baked clay cones in a central region of the site, by the Eanna ziqqurat,[1] where they formed wall decorations (Figure 1). Loftus could distinguish some motifs made with the

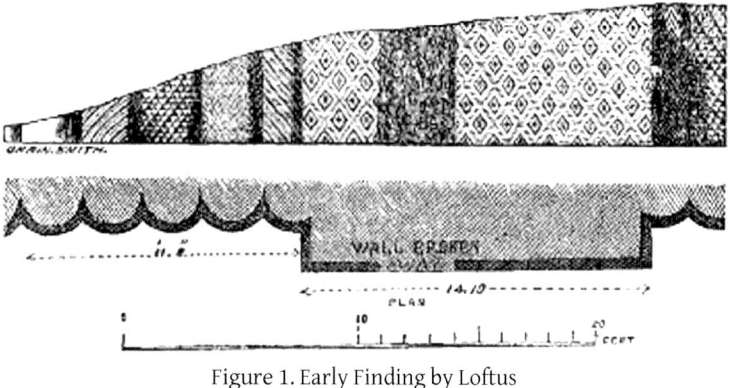

Figure 1. Early Finding by Loftus
(Loftus 1857: 188).

cones, as well as the pigments – white, red, and black – that had been used to colour their heads, and he drew parallels with archaic Egyptian funerary cones, reporting that 'no marks or inscriptions occur on these Warka cones'.[2]

Today, it seems very difficult to agree with those parallels, but during more recent research activities in Egypt, some other very interesting finds emerged which can be partly interpreted as cultural products comparable to those specimens from Uruk. Their shape and use are actually quite similar, but finds which could indicate a true affinity with Uruk seem to be still too infrequent to allow a thorough evaluation of their archaeological and historical meaning. One example of this pertains to the exploration of Thebes and

---

[1] Loftus 1857: 187–188.

[2] Loftus 1857: 189.

Figure 2. Left: Mosaics in situ at Tell 'Uqayr (Lloyd *et al.* 1943: pl. 8); top-right: Mosaics in situ by the Staircase Podium of the *Rundpfeilerhalle* at Uruk (van Ess 2013: 128); bottom-right: *Bieberschwänze* and plaster of the *Steinstiftgebäude* (van Ess 2013: 128).

comes from contexts dated to the 11th Dynasty. It was mentioned by Julius Jordan as early as the beginning of the 1930s, as it had already been related to the much later Egyptian funerary cones by Herbert E. Winlock.[3]

A second example, from Tell el-Fara'in/Buto, is a more recent discovery, but the number of specimens and their dimensions are small and their state of preservation poor.[4] Nevertheless, some parallels between the Buto findings and the Uruk cone mosaics can be reasonably proposed as their shape and stratigraphic contexts fit well together. However, the structures discovered in Thebes deserve to be interpreted entirely within the Egyptian tradition.[5]

On the other hand, the spread and the use of wall cone mosaics during the early historic period is revealed through finds covering a wide geographic area. Besides Uruk, wall cones have been found in a number of other contemporary settlements, some of which are usually connected to the diffusion of the Uruk culture during this period. A remarkable number of cones come from Southern Mesopotamian sites like Tell 'Uqayr, and various specimens were found in variable quantities in Khuzestan, South-Eastern Turkey, Northern Mesopotamia, and Northern Syria.[6]

Until now, however, it is only in the site of Uruk that such cones have been found in very large quantities and in situ in their planned architectural positions and functions. At Tell el-'Uqayr, cones with the hollows on their heads (the so-called *Grubenkopfstifte*) were also found in situ,[7] though on a much smaller scale. Therefore, it is largely due to the findings at Uruk that we can understand how the cone mosaics were used, and in which contexts (Figure 2).

From Uruk, we know that the mosaics were usually placed on the external walls and pillars of buildings that had a notable public function.[8] Sometimes, and probably during a later phase of use, mosaics were combined with friezes, which were also made of single elements combined to form large figurative representations, or with other types of works such as painted surfaces or single protruding elements.[9] According to the excavators' reconstructions, friezes were usually located on the upper part of the wall, and made a kind of frame for the system of motifs depicted on the facades. The cones with a hollow on the head were probably used as pinning supports for them, and were normally larger than the other cones.[10]

---

[3] Jordan 1931: 16; Winlock 1928: 6.
[4] See von der Way 1987: 247–249; van Ess 2011/2013: 186.
[5] For a short outline of the different interpretations of the cones in Egypt and their origins, see Joffe 2000: 113, n. 3.
[6] See for instance: Mallowan 1947: 96, pl. III; Schmidt 1978; Safar, Mustafa and Lloyd 1981: 240–241; Trokay 1981; Behm-Blanke 1989; Stein 1999: 16–18; van Ess 2011/2013: 185–186; van Ess 2013.
[7] See Lloyd 1943: 143, pl. VIII.
[8] One exception to this is, for instance, the *Pfeilerhalle*, where mosaics were also present on the inner facades of the building and in the doorways. See Brandes 1966 and 1968 for the accurate description of the types and relative position of the mosaics in the niches.
[9] Such as rosettes, representations of animals, and so on, Jordan 1931: 33–40; Heinrich 1932: 20–21.
[10] Jordan 1931: 39; van Ess 2011/2013: 184.

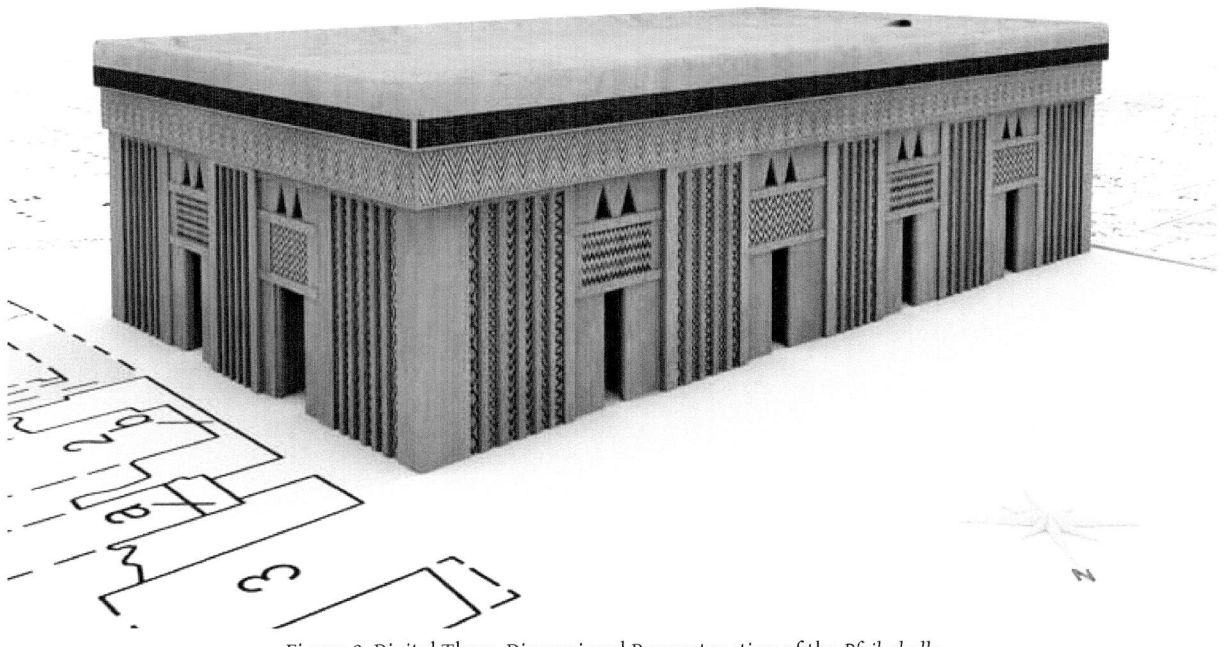

Figure 3. Digital Three-Dimensional Reconstruction of the *Pfeilerhalle*
(Eichmann 2013: 124).

Uruk

At Uruk, the cones were a structural part of the walls of some monumental buildings, so that these showed continuous motifs in different colours. Of the more imposing structures decorated in such a way are the ancient terraces, the inner sides of the enclosure wall of the *Steingebäude*, the *Rundpfeilerterrasse*, the *Pfeilerhalle* (Figure 3), and the *Steinstiftgebäude*.[11] With the exclusion of the facades of the *Steinstiftgebäude* (the mosaics of which used the natural colours of the relevant stones), the polychromatic clay cones were either made through the addition of pigments to their ends, or through different degrees of baking which rendered the clay different colours. Some proposed interpretations of the motifs depicted in the Uruk wall mosaics have been made, though there is still much to investigate regarding symbolism, as well as the uses of some of the buildings to which they belong.[12] It is, however, quite clear that the mosaics were conceived, planned and executed in tandem with their supporting architecture.[13] The clay cones were hand-made and mostly shaped in a quite rough and inaccurate way, except for their heads, that is the parts of them that had to be left visible after the plan was completely executed, which were quite accurately prepared. They were then placed in the wet plaster coat so that their heads appeared very close to each other, once the work was finished, and the thickness of the plaster coat quite often corresponded to the length of the cones (Figure 4).[14] On the walls of the buildings that were to be decorated with the mosaics, the distances between the bricks were sometimes larger than normal, perhaps to allow a better placement of the plaster and the mosaic components.[15] In the case of the *Steinstiftgebäude*, the walls contained no bricks: they were made of a kind of concrete consisting of a mix of clay, mortar and potsherds.[16] In this structure, the so-called *Bieberschwänze* (that is, 'beaver's tails') were flat ceramic elements used to block and consolidate the external side of the wall, so that the white, pinkish and dark-grey stone cones could be enough firmly fixed (see Figure 2, bottom-right). As before, the concrete cast, ceramic *Bieberschwänze*, and mosaics were all constructed at once.

Techniques and Logics

The techniques involved in the construction of the wall cone mosaics required refined skills. Above all, the use and conception of such a technique are based

---

[11] Eichmann 2007: 90, 102, 109–126, 165–168, 225–229, 234, 328, 371–385, 507; 2013, 124–126; van Ess 2015: 462–465.
[12] Some suggestions on this topic were discussed in Brandes 1968: 124–164; Behm-Blanke and Hübner 1978.
[13] The placing of the cones step by step and parallel to the progressive placement of the building materials of the architectural structures has been discussed in Brandes 1966: 17–18 (in particular page 18, where Brandes writes: 'Man wird also zu der Vorstellung kommen, daß Maurer und Mosaikleger nebeneinander und gleichzeitig auf den immer höher anwachsenden Pfeilerstümpfen arbeiteten, daß mit jeder neuen Lehmziegelschicht auch die Mosaiken in den Nischen auf die jeweils neue Höhe gebracht wurden'). See also Jordan 1931: 40; Heinrich 1932: 14; van Ess 2013: 128–129.

[14] Jordan 1931: 15–16; Heinrich 1932: 14, pl. 7; Brandes 1966: pls. 26–27.
[15] Jordan 1931: 40.
[16] van Ess 2013: 128–129.

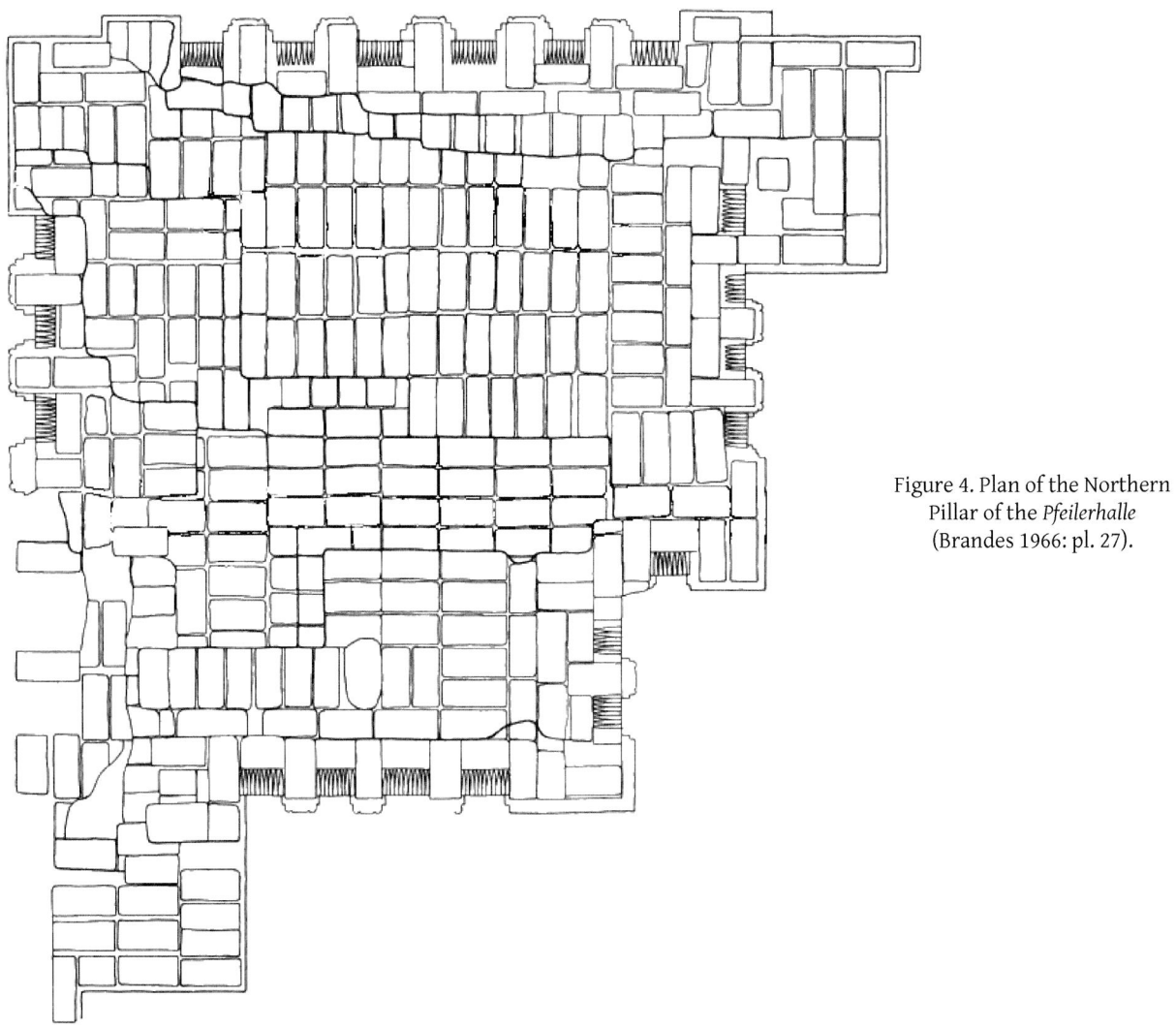

Figure 4. Plan of the Northern Pillar of the *Pfeilerhalle* (Brandes 1966: pl. 27).

on specific types of mental frameworks and cultural modalities in communication and perception. Each mosaic is actually the result of an accurate abstract project which was precisely outlined prior to the construction of the building. On the one hand, such a plan provided a coordinated parallel execution of the works: for the masonry, for the thick plaster coat of the brick structure, and for the rows of cones and other possible elements that had to be inserted in the walls.

On the other hand, that abstract project involved a monumental decoration program consisting in the *discretisation* of motifs that were made from repeated – and quite simple – compositions of lines, segments, and dots.Those responsible for the building plans and the relevant construction were thus able to master an ability which is primarily mental and logical, before being practical. It is the ability to deconstruct continuous motifs into minimal common features that can be translated into basic units that all have – more or less – the same shape and dimension. They then used these basic units to compose a large number of different motifs on the walls, possibly including novel ones that had not yet been depicted.[17] The procedures of planning the compositions and the relevant techniques might seem quite simple to a postmodern observer, but in fact, they can only be devised and outlined within specific mental frameworks which are relatively familiar with practices bound to a considerable degree of abstraction and discretisation. They are thus not easily assumed in societies that do not make extensive use of writing techniques.

Differently to a rather abstract representation, a tendentially analogical one would be based on a formal or material imitation of the model (be it concrete or ideal): for example, through the search for as close as possible an affinity. In such a case, elements of different natures or shapes would mostly be represented using

---

[17] See, for instance, for the *Pfeilerhalle*, the explanation in Brandes 1968: 9–11 (classifying the possible concrete uses of cones in the mosaic arrangements), 14–106 (the variety of recorded motives). More generally, see Eichmann 2007: 166–168.

different signs. On the other hand, in a mainly abstract communication approach, the total number of available signs is drastically reduced to few basic elements which can be combined with each other to express a large number of messages. One could think, for instance, of the logic of alphabet, or of the level reached in the realm of information technology, where everything, from images to sound to written words, can be reduced to binary digits which have all an identical nature and structure.[18]

In the case of the early historic wall mosaics, the basic unit used to draw the composition is in fact one: the round head of the baked clay or stone cone. Potential variations are introduced through the use of the different colours, and just exceptionally, or by the borders, through other elements like rosettes or friezes. Such an ability of discretisation, transforming continuous motives into assemblages of minimal units, all of which reproduce the same shape, distinctly recalls the underlying logic of present-day audio digital recording techniques, in particular those developed for master compact discs. Those masters are based on the use of samples, more precisely on the recording of large quantities of juxtaposed isolated sampled sounds. Some graphics can help to explain the differences between a digital and an analogic approach to sound recording and perception (especially the most analogic: live music played using traditional acoustic instruments – Figure 5). Digital audio recording is based on samples and could thus be compared to a line of dots which are so numerous and dense that they give the observer the impression of a single continuous mark. One second of music recorded on a compact disc is actually made of a juxtaposition of 44,100 isolated samplings, which give the listener the impression of continuous sound.[19] The 'dot-like' marks represented by the visible heads of the wall cones can thus be compared to samples ideally abstracted from a continuous representation; from a middle to long distance, they were probably not perceived as isolated heads of cones, but rather as continuous marks (to form lines and other geometric shapes).[20] The conception of a decorative program of this kind, based on the use of such cone-systems, becomes a clear indicator of the existence of a capacity for developing a relatively high-level abstract thought. In Lower Mesopotamia, the latter is probably a fundamental *novelty*, and is one of the most distinctive features of the early historic period (from the viewpoint of culturally shaped cognition and perception). The development of the first historical societies, living in urban contexts, stratified and devising the earliest forms of writing systems, is necessarily the cultural outcome of a logical passage which brings perception and communication phenomena to an increased habit to abstract. Not only writing: features like work specialisation, with a centralised bureaucracy and the development of social classes cannot come into existence without a decisive cognitive drift in that direction.

The monumental wall cone mosaics used in the early historic period in Uruk and other settlements show features that bear typical marks of the phenomena related to the first urbanisation process. The period saw not just the emergence of some types of monumental architecture, but also the birth of the relatively large settlements which hosted them: the cities.

The debate about the earliest historical societies and the emergence of the first cities is long and complex, and still fascinates and challenges scholars in various fields, but there is a general consensus that the simultaneous development of some phenomena is shared. The first forms of written accounts, work specialisation, and emerging centralised authorities appeared together with changes in technologies for agricultural production (and the subsequent economic repercussions), the development of a socio-political elite, the emergence of centralised systems to manage many aspects of community life (including cults), and the increased degree of depersonalisation in managing administration. These, and other phenomena, are all indicative of the beginning of history. They were hypothesised or reconstructed, and debated through evidence from archaeological contexts, landscape features, paleo-ecological remains, art historical findings, epigraphy, the study of artefacts, and so on. And such clues tend to converge towards a quite uniform framework. Thus, alongside some iconographies in the visual arts, the earliest administrative tablets, specific ceramic production, and so on, the wall cone mosaics provide a further clear example of changes in the mental and cultural approaches to visual representation and perception. The techniques and logic underpinning the wall mosaics, in particular the use of relatively standardised and (so to speak) 'sampled' basic visual signals to reproduce different motifs, are actually expressions and symptom of a cultural disposition which fits well with the development of early complex societies and of urbanism. Therefore, these compositions become meaningful for the interpretation of early Mesopotamian cultural history, and can aid in further broader and deeper investigations involving other contemporary evidence.

---

[18] Rambaldi 1977; Curnow and Curran 1987: 15–45, 90–102; see also the discussion in Di Ludovico 2014: 483–485.

[19] See for the underlying logics and for technical details, Watkinson 1994: 28–37. An easy popularisation of this theme can be found in Negroponte 1995: 14–17. Besides the popularising approach of Negroponte, his general perspective on communication and coding cannot be followed, as it does not properly take into consideration their logical aspects, but here they fit in quite well to give an idea of what is being dealt with.

[20] To appreciate this point of view the digital reconstructions proposed in Crüsemann should be considered: Crüsemann *et al.* 2013: 124, 128.

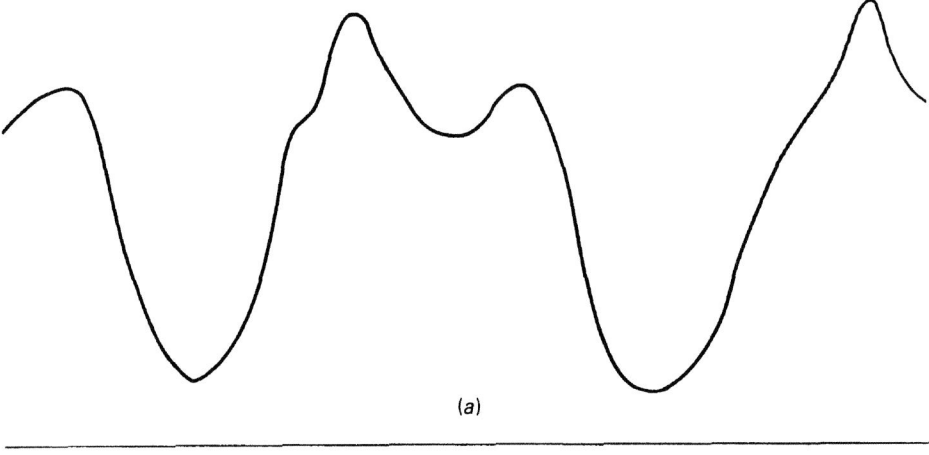
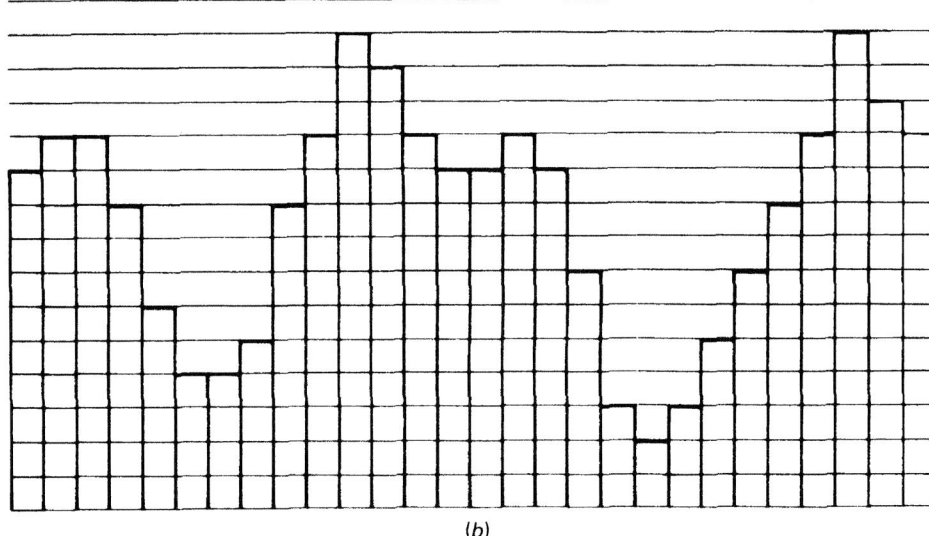

Figure 5. Example of Analogic Sound Wave and its Digital Conversion (Curnow and Curran 1987: 100).

The tendency to abstraction drives the habits of the members of the socio-cultural group towards depersonalisation and mediation in interpersonal relations, perception, and communication in general. This is also mirrored in phenomena such as the use of the cone mosaics, in which the shapes and compositions to be represented are – so to say – deconstructed in the same repeated and depersonalised form: the round cone head, whose virtual use in visual language is, according to logic, boundless. Part of this potential is well represented by the findings at Uruk and in contemporary phases of other sites.

In a recent contribution, it was proposed that the use of wall cone mosaics could be evidence that an advanced degree of the exploitation of the physiological mechanisms of visual perception had been reached.[21] In essence, this interpretation of the phenomenon of the Late Uruk mosaics refers to the mechanisms of vision, as they are currently explained in medical disciplines, and their neurological backgrounds. It is therefore suggested that the mosaics represent a concrete summary of the way visual perception works. It demonstrates the decomposition of an image to minimal points, its reproduction through the enhancing, if needed, of chromatic contrasts, and its reduction to basic modules (or motifs) for translation and transmission, so that the image could be regularly reproduced in other places and by other peoples. The authors even hypothesise the existence of pictographic written messages which would synthesise the compositional scheme, so that it could be communicated over a great distance, and the mosaic could be accurately reproduced elsewhere. Such an interpretation cannot be accepted using the perspective adopted here, since it projects postmodern logic onto the Early Historic technique of wall cone use. On the contrary, discretisation in coding strategies and the complex visual communication issues connected to it deserve to be investigated properly. It is in fact too easy to adopt, more or less consciously, approaches which lead to symmetrisation, to misunderstandings caused by violations of the logical hierarchies and, so to speak, illusions in the perspective. Misunderstandings originate, in the case of the contribution by Roi and Girard, from three main violations.

---

[21] Roi and Girard 2013. The existence of this work only became known to the author recently, while preparing the draft of this paper for publication.

Firstly, the authors interpret Mesopotamian proto-historical phenomena through implicit perceptive schemes which belong to a few postmodern cultures, without considering any feature in their model that would allow for the appreciation of the remarkable distances and differences between cultural contexts. The second main logical violation, connected to and supporting the first, lies in the assumption that proto-historical human perception mechanisms would be based on the same logic and mode of operation as postmodern ones. This cannot be proved; on the contrary, there is rather evidence which permits us to reject such view, since spatial perception, visual perception, and the general use of the senses are deeply shaped by the culture to which one belongs.[22] The latter are, in any case, so closely connected with communication and perception processes that their existence and the way they work cannot be easily extended to different historical periods and cultures. However, some general features common to different contexts can be detected, and some irreducible differences as well. The third great source for misunderstandings is the lack of consideration for how our interpretation and understanding of the perceptive mechanisms is itself conditioned by cultural factors, even in supposedly objective medical laboratory observation.[23]

An analysis such as the one presented here, paralleling the Uruk mosaics and digital audio coding, can be only used as a comparison of distant phenomena which share a few general basic principles working in similar ways, and as the detection of a mechanism which is partially common to both coding systems. The difference is essential, since it is of great methodological importance. The actual logic, and the cultural and historical contexts to which each of these coding systems belong, are differentiated by a number of important fundamental elements.

The methodology employed by this contribution needs to be expanded, and other issues linked to the wall mosaics should be properly investigated in the light of what has been discussed here: of prime importance, the cultural and social meanings and values of the depicted motifs, as well as their role within the more general historical framework in which Late Uruk society and cultures developed.

Acknowledgments

The author warmly thanks Ch. Purschwitz, S. Bracci, and N. Morello for the help in finding some of the bibliographic materials.

---

[22] Le Breton 2006.
[23] The issues and topics recalled here have been discussed by the author in existing works, see, for example, Di Ludovico 2014.

References

Behm-Blanke, M.R. 1989. Mosaikstifte am oberen Euphrat: Wandschmuck aus der Uruk-Zeit, *Istanbuler Mitteilungen* 39: 73–83.

Behm-Blanke, M.R. and W. Hübner 1978. Ein frühsumerisches Kalenderhaus in Uruk-Warka?, *Baghdader Mitteilungen* 9: 134–136.

Brandes, M. 1966. Beobachtungen zu den Stiftmosaiken der Pfeilerhalle Iva. In H.J. Lenzen, *XXII. vorläufiger Bericht über die von dem deutschen archäologischen Institut und der Deutschen Orient-Gesellschaft aus Mitteln der Deutschen Forschungsgemeinschaft unternommenen Ausgrabungen in Uruk-Warka*: 15–20. Berlin: Gebr. Mann. Verlag.

Brandes, M. 1968. *Untersuchungen zur Komposition der Stiftmosaiken an der Pfeilerhalle der Schicht IVa in Uruk-Warka*, Baghdader Mitteilungen. Beiheft 1. Berlin: Gebr. Mann. Verlag.

Crüsemann, N., M. van Ess, M. Hilgert and B. Salje (eds) 2013. *Uruk. 5000 Jahre Megacity. Begleitband zur Ausstellung 'URUK – 5000 Jahre Megacity' im Pergamonmuseum - Staatliche Museen zu Berlin, in Den Reiss-Engelhorn-Museen Mannheim*. Petersberg: Michael Imhof Verlag.

Curnow, R. and S. Curran 1987. *Il primo libro di informatica*. Torino: Boringhieri Editore.

Di Ludovico, A. 2014. The Reign of Šulgi. Investigation of a King above Suspicion. In H. Neumann, R. Dittmann, S. Paulus, G. Neumann and A. Schuster-Brandis (eds), *Krieg und Frieden im Alten Vorderasien. Proceedings of the 52e Rencontre Assyriologique Internationale, Münster, 17.-21. Juli 2006*. Alter Orient und Altes Testament 401: 481-493. Münster: Ugarit-Verlag.

Eichmann, R. 2007. *Uruk. Architektur I. Von den Anfängen bis zur frühdynastischen Zeit,* Ausgrabungen in Uruk-Warka Endberichte 14. Rahden, Westf.: Verlag Marie Leidorf.

Eichmann, R. 2013. Frühe Großarchitektur der Stadt Uruk. In N. Crüsemann *et al.* (eds) 2013: 117–127.

van Ess, M. 2011/2013. Stiftmosaik, *Reallexikon der Assyriologie und Vorderasiatischen Archäologie* 13, Berlin 2011/2013: 184–186.

van Ess, M. 2013. Die Technik der Tonstiftmosaiken. In N. Crüsemann *et al.* (eds) 2013: 128–129.

van Ess, M. 2015. Uruk. B. Archäologisch, *Reallexikon der Assyriologie und Vorderasiatischen Archäologie* 14, 5./6.: 457–487.

Heinrich, E. 1932. Die Schichten und ihre Bauten. In A. Nöldeke, *Vierter vorläufiger Bericht über die von der Deutschen Forschungsgemeinschaft in Uruk-Warka unternommenen Ausgrabungen*: 6–24. Berlin: Gebr. Mann. Verlag.

Joffe, A.H. 2000. Egypt and Syro-Mesopotamia in the 4th Millennium: Implications of the New Chronology, *Current Anthropology* 41: 113–123.

Jordan, J. 1930. *Erster vorläufiger Bericht über die von der Notgemeinschaft der Deutschen Wissenschaft in Uruk* unternommenen Ausgrabungen. Berlin: Verlag der Akademie du *Wissenschaften.*

Jordan, J. 1931. *Zweiter vorläufiger Bericht über die von der Notgemeinschaft der Deutschen Wissenschaft in Uruk unternommenen Ausgrabungen.* Berlin: De Gruyter.

Le Breton, D. 2006. *Le Saveur du Monde. Une anthropologie des sens.* Paris: Métaillé.

Lloyd, S., F. Safar and H. Frankfort 1943. Tell Uqair: Excavations by the Iraq Government Directorate of Antiquities in 1940 and 1941, *Journal of Near Eastern Studies* 2: 131–158.

Lenzen, H.J. 1966. *XXII. vorläufiger Bericht über die von dem Deutschen Archäologischen Institut und der Deutschen Orient-Gesellschaft aus Mitteln der Deutschen Forschungsgemeinschaft unternommenen Ausgrabungen in Uruk-Warka.* Berlin: Gebr. Mann.

Loftus, W.K. 1857. *Travels and Researches in Chaldea and Susiana; with an Account of Excavations at Warka, the 'Erech' of Nimrod, and Shush, 'Shushan the Palace' of Esther, in 1849-52.* New York: Robert Carter and Brothers.

Mallowan, M.E.L. 1947. Excavations at Brak and Chagar Bazar, *Iraq* 9: 1–259.

Negroponte, N. 1995. *Being Digital.* London: Hodder & Stoughton.

Rambaldi, E. 1977. Astratto/concreto. In *Enciclopedia Einaudi 1977-1984*, vol. 1: 1011–1060. Torino: Einaudi.

Roi, Ph. and T. Girard 2013. L'image de cônes et le système visuel. *La Théorie Sensorielle. I - Les Analogies Sensorielles*: 117–128. Brussels, First Edition Design Publishing. Online <http://www.theoriesensorielle.com/analogie-entre-limage-de-cones-et-le-systeme-visuel/> (last access 26 Sept. 2016).

Safar, F., M.A. Mustafa and S. Lloyd 1981. *Eridu.* Baghdad: Ministry of Culture and Information, State Organization of Antiquities and Heritage.

Schmidt, J. 1978. Tell Mismar: ein prähistorischer Fundort im Südiraq, *Baghdader Mitteilungen* 9, 10–17.

Stein, G. 1999. Material Culture and Social Identity: the Evidence for a 4th Millennium BC Mesopotamian Uruk Colony at Hacinebi, Turkey, *Paléorient* 25/1: 11–22.

Taylor, J.E. 1855. Notes on Abu Shahrein and Tel el Lahm, *Journal of the Royal Asiatic Society* 15: 404–415.

Trokay, M. 1981. Les cônes d'argile du Tell Kannâs, *Syria* 58: 153–171.

Watkinson, J. 1994. *An Introduction to Digital Audio.* Oxford: Focal Press.

von der Way, T. 1987. Tell el Fara'in-Buto: 2. Bericht, *Mitteilungen des Deutschen Archäologischen Instituts Kairo* 43: 247–250.

Winlock, H.E. 1928. The Egyptian Expedition 1925-1927: the Museum's Excavations at Thebes, *The Metropolitan Museum of Art Bulletin* 23: 3–58.

Section 2

Symbols in Action

Composite animals in Mesopotamia as cultural symbols

Chikako E. Watanabe
Osaka Gakuin University, Japan
chikako@ogu.ac.jp

Abstract

Composite animals are the product of our thought processes. They are nonexistent in reality but their occurrence is observed universally. They exhibit a body structure that consists of multiple body parts taken from different animals of reality in order to form a single creature. This study focuses on a particular type of composite animal – the lion dragon – to elucidate its symbolic function through artistic and textual descriptions, and also its predecessor, the lion-headed eagle. A peculiar pose held by the creature is examined in order to reveal its specific role associated with an auditory aspect of the thunderstorm. The importance of the creature's leonine head is indicated by the application of a special material (gold) to the head of the lion-headed eagle in some artefacts manufactured with composite materials.

Keywords: lion dragon, lion-headed eagle, Anzû, imdugud, storm god, thunderstorm

1. Introduction

This paper aims to examine aspects of the lion dragon, one of the oldest composite animals in Mesopotamia, from the point of view of the aetiological and symbolic functions it embodies, and its relation to the materiality of some artefacts. The predecessor of the lion dragon, the lion-headed eagle, dates back beyond the 3rd millennium. Among several important previous studies on composite animals, pioneering work was carried out by Anthony Green and Frans Wiggermann,[1] both systematically identified the names of creatures in texts with their visual representations. Wiggermann examined a wide range of ritual and religious texts, while Green investigated monsters and demons represented in art, and laid the groundwork for clarifying their nature and identity. The names of these creatures vary depending on the scholars cited, but in this study I will follow the naming used by Wiggermann and Green in *Reallexikon der Assyriologie*.[2]

1.1. Previous Studies on Materiality

There have been extensive discussions in the context of recent archaeological theories about what 'materiality' is, in which scholars' views of the term appear quite diverse and divided; there is no simple and straightforward definition of the term. According to C. Knappett, when the term 'materiality' is used, it describes a 'phenomenon' which is to be regarded as 'relational' rather than the 'static' or 'categorical' concepts expressed by terms such as artefacts, materials, material cultures or the material world.[3] He claims that the term is intended to express the ongoing process/dynamics of human-artefactual relations, and the fundamental aim of such research is to reduce the duality between mind and matter. Knappett defines the term as covering four different areas or concepts: (1) dependent (material relations), (2) co-dependent (social relations), (3) independent (vital), and (4) interdependent (plural) properties of objects.

T. Ingold, however, argued that it is more productive to concentrate on the 'property' of materials and on the process of historical changes than to deal with a concept that is ambiguous. He regards materials as being 'varieties of matter' and 'of the physical constitution of the world' as a result of the presence or activity of its inhabitants. Hence such properties are seen as properties of matter, and are opposed to the qualities that the mind imaginatively projects onto them. Ingold chose to concentrate on 'substances' and 'media' by following J.J. Gibson, and the 'surfaces' between them which are regarded as the basic components of the 'environment', and not the physical or material world. Ingold explains the environment as 'a world that continually *unfolds* in relation to the beings that make a living there. Its reality is not *of* material objects but *for* its inhabitants'.[4] According to Ingold, it is a world of 'materials' which occur as the environment unfolds. Thus the properties of materials are regarded as constituents of an environment, and they are not fixed attributes of the items made from those materials, but are rather processual and relational. Ingold's view was

---

[1] Green 1983; 1986; Wiggermann 1989; 1992.
[2] Wiggermann and Green 1994: 222-264.
[3] Knappett 2014: 4701-2.
[4] Ingold 2007: 14. Cf. Gibson 1979: 8.

later criticized for risking the separation of materiality from its social context,[5] but his argument has raised an important question and needs to be considered in the discussion of materiality and material engagement.[6]

In the field of ancient Near Eastern studies, D. Wengrow emphasised the materiality of sacred power, in which the function of temple construction was reinterpreted in the context of the king's domination by localising a god within its temple.[7] B. Pongratz-Leisten examined the composite nature of the body and divine agency by analysing religious and mythological texts with a special focus on a combat myth of the god Ninurta.[8] K. Sonik studied the Mesopotamian divine image to analyse the nature of the relationship between the image and its divine significance.[9] The latter two authors emphasised the composite nature of the divine entities featured by their interaction with aspects of materiality.

This paper aims to elucidate a specific function of the lion dragon/lion-headed eagle in embodying and making visible the auditory aspect of the thunderstorm. It is intended to examine not only the creature's composite body structure but also the use of composite materials in manufacturing artefacts of this creature.

2. The Lion-Headed Eagle

The lion-headed eagle, which comprises a bird of prey with the head of a lion, appears in the earliest pictorial representations shown in seal impressions which date back to the Uruk period.[10] In this early period, the creature is represented in profile flying over captured enemies with wings stretched upright and head lowered; occasionally something is shown being spat out of its open mouth onto the ground.[11] From the Early Dynastic to Akkadian periods the motif of the lion-headed eagle appeared frequently.[12]

Different types of stones and metals were used, and precious materials, such as gold, silver and lapis lazuli, were sometimes employed in the manufacture of artefacts of the creature. During the Early Dynastic period the lion-headed eagle was depicted in frontal view with wings and legs spread wide to stand over a pair of animals, such as ibexes, stags or lions. F. Wiggermann identified the lion-headed eagle as Anzû and, when the creature is combined with the pairs of animals, they were thought to be associated with the god Enki in the case of the ibex, the goddess Ninhursag in the case of the stag, and the god Ningirsu in the case of the lion.[13] The creature is also depicted on the 'Stele of the Vultures' together with a pair of lions' heads, which are represented below the lion-headed eagle, on top of a net. The net contains naked enemies of Girsu; a large male figure grasps the tail feathers of the lion-headed eagle.[14]

A famous example of the creature comes from Mari in the Middle Euphrates region[15] where the creature is represented in a frontal view with the wings spread wide; its body and wings are incised in flat lapis lazuli, and the head, neck and tail are fashioned in bitumen covered with gold leaf. The sculpture was found buried under the floor of the Pre-Sargonic palace, having being placed inside a sealed clay jar full of jewellery, cylinder seals and statuettes of Sumerian origin (the so-called 'treasure of Ur'). In the context of composite materiality, the application of different materials to different body parts of the creature is worth noting, as it may indicate the significance of material engagement to emphasise a particular function and notion perceived in a specific body part.

There are two other examples comparable to the Mari artefact. The first comes from Tell Brak in the Habur area. The creature is represented in a frontal view and not only its wings but also its legs are spread wide with both feet flexed upwards.[16] The creature's body is carved in lapis lazuli while its head is made from a gold-sheet mask covering a core of frit (a fused or partially fused ceramic composition). This piece was discovered in a room of an Akkadian-period building in the area HS and, like the Mari example, it is believed to have been manufactured in southern Mesopotamia and brought to Tell Brak as a personal possession during the Early Dynastic period. Although the tail is not covered in gold leaf, the application of gold to the head, together with the lapis lazuli body, makes it similar to the Mari representation. It should be noted that the head is treated specially in both instances by using an elaborate technique of gold-sheet application, which makes the creature's head radiant and stand out against the dark blue body and wings. The importance of the creature's head in the context of thunderstorms will be discussed in the following section. The second example comes from Tell Asmar in the Diyala region,

---

[5] Knappett 2007; Miller 2007; Tilley 2007.
[6] Renfrew 2004; Malafouris and Renfrew 2010.
[7] Wengrow 2004.
[8] Pongratz-Leisten 2015.
[9] Sonik 2015.
[10] Amiet 1980: fig. 1602, pl. 120; Brandes 1979: pl. 12.
[11] Amiet 1980: pl. 13.
[12] Braun-Holzinger 1990: 94-97.

[13] The lion-headed eagle represents a superior power, suggesting possibly the power of the god Enlil. See Wiggermann and Green 1994: 226 (§ 2.1).
[14] There is a problem here, however, if we take Wiggermann's interpretation of the lion-headed eagle as representing the power of the supreme god, Enlil, since his emblem is grabbed by the left hand of an anthropomorphic male figure. Irene Winter identified this figure as the god Ningirsu, the city god of Lagash (Winter 1985: 13-15). The representation of Ningirsu grasping the tail of a superior god in the pantheon may not be appropriate.
[15] Orthmann 1975: fig. 122a; Moortgat and Moortgat-Correns 1974: fig. 4.
[16] Matthews 1994: 294, figs. 7-9.

where three small figures of the lion-headed eagle were included in the late Akkadian hoard.[17] One figure has a body of lapis lazuli with silver wings, head and tail, while the other two have lapis lazuli wings and silver torsos, heads and tails. In these examples, lapis lazuli is applied to different body parts, whereas the creature's head and tail are consistently silver.

Lapis lazuli is a semi-precious stone highly praised for its beauty.[18] Textual sources of the 3rd millennium indicate that the stone reached Mesopotamia from the east via regions such as Aratta, Dilmun and Meluhha.[19] It is believed that the source of the lapis lazuli brought to Mesopotamia was on the upper reaches of the Kokcha river, a tributary of the Oxus, in the Badakhshan district of modern Afghanistan.[20]

3. Thunderstorms and the Lion Dragon

The depiction of the lion-headed eagle in frontal view, which was so common in the Early Dynastic period, disappears gradually – instead the creature starts to be represented in profile, standing on four feet with its forepaws those of a lion and its hind legs and tail those of a bird of prey.[21] This composite creature is called 'lion dragon'.[22] The iconographic transition is clearly illustrated on an incised shell plaque,[23] dated to the late Early Dynastic period: the creature is depicted in frontal view and in profile in the upper and lower registers of the plaque respectively. In the following Akkadian period, the creature no longer appears alone, but is shown with an anthropomorphic god and a female figure in a scene represented on cylinder seals.[24] The god rides on a chariot with a whip in his hand; the whip is known as an attribute of the storm god Iškur/Adad from textual evidence.[25] His chariot is pulled by a lion dragon, which takes the characteristic posture of lowering its head towards the ground with its mouth wide open, from which two wavy lines extend. There is a worshipper or priest performing a ritual of libation before an altar that is set up in front of the lion dragon. The female figure appears above the lion dragon with wavy lines in her hands, which may represent rainfall. The liquid poured from the vessel held by the worshipper is depicted by a similar wavy line. The scene as a whole seems to represent the natural phenomenon of a storm with the god Iškur/Adad in the centre of the scene.

A question arises: what is the lion dragon spitting out? Although it looks as though the creature may be vomiting, should the substance in question be regarded as a liquid, such as representing rain? I would like to seek an answer to the meaning of this scene from the creature's name and its peculiar body posture. Both Wiggermann and Green identify the lion-headed eagle as the original form of Anzû.[26] The reading of the name of Imdugud/Anzû was much debated by scholars such as B. Landsberger, W.G. Lambert and B. Alster,[27] however, the philological argument around the name is not dealt with here. Instead I would like to focus on the basic meaning of the name of the creature. The name consists of four signs: AN. IM. DUGUD. and MUŠEN meaning literally 'the bird (mušen) of heavy cloud/fog (im.dugud = imbaru) in the sky (an)', which suggests a close association with thick cloud. The Akkadian lion dragon associated with the storm god Iškur/Adad is identified by Wiggermann as u_4.ka.duḫ.a in Sumerian and ūmu nā'iru in Akkadian, which is translated as 'Roaring Day' to embody turbulent weather. The Sumerian name u_4.ka.duḫ.a literally means 'the day of wide-open mouth', and in lexical text, the Akkadian words nā'iri and kaduḫḫû are equated with šegû meaning 'raging'. This name therefore suggests the roar of the creature in association with the storm god – in other words, the lion dragon provides the raging sound of 'thunder' produced by its 'wide-open mouth (ka.duḫ.a)'.[28]

In the representation of Akkadian cylinder seals, we should look carefully at the posture of the lion dragon, because features associated with the sight of heavy clouds as well as the auditory aspect of thunder can be explained from the peculiar posture of the creature. Both the lion-headed eagle and the lion dragon have their heads lowered towards the ground with their mouths wide open. In fact, this posture is typically taken by real lions when they roar. It is explained in *Grzimek's Animal Life Encyclopedia* as follows: 'when they roar, lions usually stand and tilt the head slightly toward the ground. The flanks are drawn in, and the chest expands like a mighty balloon. Dust is often stirred up in clouds as the lion roars, so forcefully is the air expelled'.[29]

Lions roar most mightily shortly after sunset for a period of about an hour, and B. Grzimek described the sound of the roar as 'the most magnificent sound in all creation'. A lion's roar is so loud that it can be heard at a distance of 8 to 9 km. It is noteworthy that clouds of dust have been observed when lions roar, due to the exhaled air which stirs dust particles up into the air. Thus it is likely that, with the wide-open mouth of a lion, Anzû

---

[17] Marchetti 1996; Moortgat and Moortgat-Correns 1974: fig. 5a; Matthews 1994: 296.
[18] Winter 1999.
[19] Moorey 1999: 85–92.
[20] Moorey 1999: 86.
[21] For a study on the lion-headed eagle and the lion dragon in association with the storm god, see Green 2003: 24–34.
[22] Braun-Holzinger 1990: 97–99.
[23] Parrot 1948: 113, fig. 27m; Amiet 1980: fig. 1278.
[24] Frankfort 1939: pl. XXIId; Buchanan 1966: fig. 335; Collon 1982: figs. 137 and 192; Collon 1987: fig. 779.
[25] Tallqvist 1938: 246–249.

[26] Wiggermann and Green 1994: 225–226 (§ 2.1), 243 (§7.14), 258 (§3.25).
[27] Landsberger 1961; Lambert 1980: 81–82; Alster 1991: 4–5.
[28] Watanabe 2002: 93–98.
[29] Grzimek 1972: 357.

roars in order to produce the reverberating noise of thunder in the sky, which may be complemented by the rumbling noise of the chariot driven by the storm god. In addition, the thick clouds of a thunderstorm were probably perceived as being formed by the strongly expelled breath of the creature, in the same way real lions stir up dust clouds when they roar. Anzû therefore is responsible for the storm clouds (im.dugud) and the noise of thunder when it roars with the wide-open mouth (ka.duḫ.a) of a lion, and this, of course, should take place in the sky (an), which is made possible by the eagle's wings and tail. This scene can be regarded as a clear example of the composite animal serving to provide aetiological explanations of the natural phenomenon of the thunderstorm. The role of the creature is closely associated with the divine power represented by the anthropomorphic god, in which the creature embodies a concrete part of the phenomena controlled by the storm god.

4. The Lion Dragon as an Enemy of Ninurta

From the beginning of the 2nd millennium, the storm god is shown more closely associated with another of his animal attributes, the bull. The lion dragon represents Anzû independently, and was at first depicted as a faithful divine servant, as described in the epic of Lugalbanda, in which Anzû makes the clouds dense and roars at the rising sun; the creature blocks enemy forces at the command of Enlil.[30] In Gudea Cylinder A, Anzû is still described as a divine emblem in close association with the god Ningirsu,[31] who is a local form of the divine hero Ninurta in the City State of Lagash. However, some time during the Ur III period, the role of Anzû changed, and the creature is suddenly counted among the slain enemies of the god Ninurta.[32] In the myth Lugal-e, which deals with the victory of Ninurta over Asakku,[33] and another myth Angim, which deals with Ninurta's victorious return to Nippur from an expedition to the mountains,[34] Anzû appears as one of the eleven monsters slain by the god. Hence Anzû plays a role predominantly as a wicked monster which stole the Tablet of Destinies from its righteous divine owner, and for this evil deed the creature is to be vanquished by the divine hero Ninurta. The adoption of Anzû as Ninurta's animal attribute is explained in the myth Ninurta and the turtle, in which the god Enki tells Ninurta that he will place his feet on Anzû's neck, since the divine hero caught the creature with his mighty weapon.[35]

In the 1st millennium, representations of Anzû can be divided into two groups: firstly, Anzû is pursued by the god Ninurta, and secondly, Anzû is depicted as a divine attribute with an anthropomorphic god standing on its back. The importance of the god Ninurta increased in the 9th century when Ninurta became the city-god of the new Assyrian capital Kalhu (Nimrud). An iconographic representation of the deity was apparently sought urgently, because no divine image of Ninurta existed in Assyria. Ashurnasirpal II, the founder of the new capital, writes in his annals:[36] 'At that time I created upon my own intuition this statue of the god Ninurta which had not existed previously as an icon of his great divinity out of the best stone of the mountain and red gold. I regarded it as my great divinity in the city Calah'.[37] The statue mentioned here is probably a divine statue placed in his temple which has not survived. However, a stone relief depicting the deity was discovered in Kalhu in the ruins of the Ninurta Temple, which shows Ninurta confronting Anzû.[38] This stone slab once flanked the entrance to the Ninurta Temple and bears a text on the surface which describes dedication to the deity.[39] The relief shows a winged god, holding three-pronged lightning bolts in both hands, running after Anzû. The creature's head is turned back towards the deity with open mouth; its head and forepaws are those of a lion, the tail, wings and hind legs are those of an eagle, and its body is covered with feathers. If the claim of Ashurnasirpal to have created Ninurta's image 'upon his own intuition' (ḫissat libbīja)[40] was true, the depiction of Ninurta on a bas relief from the Ninurta Temple would have been one of the earliest examples of the deity in this period.

The theme of Ninurta chasing Anzû was repeated as a common motif on seal impressions with a curious modification. Although Anzû is depicted in the same style, the god is shown accompanied by a slightly different composite animal, the horned lion griffin, which is galloping under the god who is also shown in a running posture.[41] The horned lion griffin comprises four different animals: the creature has the head of a lion with additional bull's horns, lion's forepaws, bird's wings and hind legs, and the tail of a scorpion. This creature occurs predominantly in the Neo-Assyrian period with a few instances of its forerunner in the second half of the 2nd millennium. U. Seidl examined the representation of this creature and identified it as the flood-monster abūbu. The creature is depicted surmounted typically by the god Ninurta who aims at Anzû with a bow and arrow. The running posture of Ninurta is new and worth examining. When Anzû was conquered by the divine hero in the myth, Ninurta

---

[30] Jacobsen 1987: 327: 101.
[31] Gudea Cyl. A, XIII 22; Edzard 1997: 77.
[32] Watanabe 2002: 127–131.
[33] Jacobsen 1987: 243, 129–131.
[34] Cooper 1978: 60: 39, 64: 61.
[35] UET 6/1 2: 16ff.; Alster 1972: 120–125.

[36] Grayson 1976: 136–137: 576.
[37] Grayson 1991: 212, ii 132–4.
[38] Moortgat-Correns 1988: pl. 3.
[39] Grayson 1976: 177: 683–686.
[40] CAD vol. 6, 202 ḫissatu (1)–(c) in ḫissat libbi.
[41] Porada 1948: figs. 689, 692; Moortgat-Correns 1988: fig. 5; Frankfort 1939: pl. XXXVb.

used the storm wind to make the creature's wings droop, then shot arrows and cut off Anzû's wings.[42] No act of running is involved in the account of the myth. Ninurta's characteristic posture is, however, explained in a cultic commentary. A text called *Marduk Ordeal* mentions that the victory of Ninurta over Anzû was enacted in all cult-centres in the month of *Kislimu* as a cultic footrace called *lismu*: 'The footrace (*lismu*) which they go [round] in front of Bēl and in all the cult places in Kislim [is that of Ninurta]. When Aššur sent Ninurta to vanquish Anzû, ...'.[43] Thus, the depiction of Ninurta in a running posture was probably intended to link the god's pictorial image with his cultic footrace. We don't know whether this representation was invented by Ashurnasirpal with his intuition but it must certainly have been created during his reign.

When this mythological episode was depicted in places other than the temple of Ninurta, a difficulty must have arisen concerning how to differentiate Ninurta's enemy from his associated animal. Animal attributes function to provide a clue to identifying the deity represented in a scene. Ninurta's traditional associated animal was Anzû, as stated in the myth *Ninurta and the turtle*, however, a depiction of Anzû as the divine attribute on which the god rides would have created confusion when the god is shown chasing after the same creature. The horned lion griffin may have been introduced into this scene in order to solve this problem. When a visual representation of the god Ninurta's associated animal other than Anzû was sought, the horned lion griffin was probably a good solution, since it comprises the new factors of the bull and scorpion in addition to those of the lion and eagle which are in common with Anzû.

5. The Lion Dragon as Divine Attributes

Considering the other group of representations, in which Anzû is depicted as an animal attribute of a deity in the 1st millennium, again the depiction falls into two categories. In the first category, Anzû is associated with a goddess and, in the second, Anzû accompanies the god Adad. When Anzû has a goddess standing on its back, the scene includes a male deity who stands on his own associated animal. The goddess is shown either surrounded by an aura decorated with stars,[44] or armed with a quiver and sword.[45] The goddess with the aura appears with the god Nabû, who stands on the snake dragon, *mušḫuššu*, holding a stylus in his hand, while the armed goddess appears with a god standing on a bull who may be Adad. The identity of the goddess is not certain, but she may well be the goddess Ištar, who is identified with the planet Venus and is responsible for war. Alternatively, she may be the spouse of the male deity represented in the scene.

In Babylon, Anzû was shown associated with Adad. A lapis lazuli cylinder dated to the 9th century is engraved with the image of Adad holding Anzû by a tether – Anzû is crouching at the foot of the god together with a bull.[46] This was originally the seal of Adad, which was later dedicated to Marduk by the Assyrian king Esarhaddon, who reconstructed the city of Babylon which had been destroyed by his father Sennacherib.[47] Thus, in the 1st millennium, the figure of Anzû was associated not only with Ninurta but also with Adad and perhaps Ištar. The creature's association with both the storm god Adad and the divine hero Ninurta goes back to the 3rd millennium, but its association with a goddess is new, and I do not have a good explanation for it. Anzû's new association with the goddess, however, suggests that the creature served in this period as a neutral symbol, which could be flexibly adapted according to a combination of notions beyond its conventional role defined by the myth. The creature's mythological aspect was emphasised only when it appeared in a scene where it was being chased by the god Ninurta. It is noteworthy that the creature's original association with the thunderstorm was still alluded to by the trident held by the deity, which is often interpreted as thunderbolt, in the representation of the relief placed at the entrance to the Ninurta Temple in Kalhu.

Conclusion

From the 3rd millennium, the lion dragon was closely associated with the storm god. The creature embodied the concrete phenomena of the thunderstorm, i.e., the noise of thunder and the thick clouds, which were believed to be brought about by the creature's roar. In the 1st millennium, Anzû was shown being chased by Ninurta, reflecting a cultic footrace performed to commemorate the divine hero's victory over the creature. When the lion dragon was used as a divine attribute in the 1st millennium, Anzû was adopted beyond its original divine association, in which the creature acted as a symbol that can evoke notions readily according to the different contexts provided.

Acknowledgement

This work is dedicated to the memory of Tony (Dr. Anthony Green). The study was supported by JSPS KAKENHI Grant Number JP26283012.

References

Alster, B. 1972. Ninurta and the Turtle, UET 6/1 2, *Journal of Cuneiform Studies* 24: 120–125.
Alster, B. 1991. Contributions to the Sumerian Lexicon, *Revue d'Assyriologie et archéologie orientale* 85: 1–11.

---

[42] Vogelzang 1988.
[43] Livingstone 1986: 240–1: 57–8; 1989: 84–5, no. 34: 57–8.
[44] Porada 1948: fig. 691.
[45] Collon 1987: fig. 351; Porada 1948: fig. 694.
[46] Jakob-Rost *et al.* 1992: fig. 67; Collon 1987: fig. 563.
[47] A later inscription reads -the property of Marduk, great lord, his lord, Esarhaddon, king of the universe, king of Assyria, has given (this seal) for his life-: Collon 1987: 131 and 134, fig. 563.

Amiet, P. 1980. *La glyptique mésopotamienne archaïque*, 2nd edition. Paris: Editions du CNRS.

Brandes, M.A. 1979. *Siegelabrollungen aus den archaischen Bauschichten in Uruk-Warka*. Freiburger altorientalische Studien 3. Wiesbaden: Franz Steiner Verlag,

Braun-Holzinger, E-A. 1990. Löwenadler. In E. Ebeling and D.-O. Edzard (eds), *Reallexikon der Assyriologie und vorderasiatischen Archäologie*, Band 7: 94-97. Berlin: de Gruyter.

Buchanan, B. 1966. *Catalogue of the Ancient Near Eastern Seals in the Ashmolean Museum I. Cylinder seals*. Oxford: Clarendon Press.

Collon, D. 1982. *Catalogue of Western Asiatic Seals in the British Museum, Cylinder Seals II (Akkadian-post Akkadian-Ur III periods)*. London: British Museum.

Collon, D. 1987. *First Impressions: Cylinder Seals in the Ancient Near East*. London: British Museum.

Cooper, J.S. 1978. *The Return of Ninurta to Nippur: an-gim dím-ma*. Analecta Orientalia 52. Rome: Pontificium Institutum Biblicum.

DeMarrais, E. *et al.* 2004. E. DeMarrais, C. Gosden and C. Renfrew (eds), *Rethinking Materiality: the Engagement of Mind with the Material World*. Cambridge: McDonald Institute for Archaeological Research.

Edzard, D.O. 1997. *Gudea and his Dynasty. The Royal Inscriptions of Mesopotamia: Early Periods*, vol. 3/1. Toronto: University of Toronto Press.

Frankfort, H. 1939. *Cylinder Seals: a Documentary Essay on the Art and Religion of the Ancient Near East*, London: Macmillan.

Gibson, J.J. 1979. *The Ecological Approach to Visual Perception*. Boston: Houghton Mifflin.

Grayson, A.K. 1976. *Assyrian Royal Inscriptions. Part 2. From Tiglath-pileser I to Ashur-nasir-apli II. Records of the Ancient Near East II*. Wiesbaden: Harrassowitz.

Grayson, A.K. 1991. *Assyrian Rulers of the Early First Millennium B.C. I (1114-859 B.C.)*. The Royal Inscriptions of Mesopotamia, Assyrian Periods vol. 2. Toronto: University Press.

Green, A. 1983. Neo-Assyrian Apotropaic Figures: Figurines, Rituals and Monumental Art, with Special Reference to the Figurines from the Excavations of the British School of Archaeology in Iraq at Nimrud, *Iraq* 45: 87–96.

Green, A. 1986. The Lion-Demon in the Art of Mesopotamia and Neighbouring Regions, *Baghdader Mitteilungen* 17: 141–254.

Green, A.R.W. 2003. *The Storm-God in the Ancient Near East*. Biblical and Judaic Studies from the University of California, San Diego 8. Winona Lake: Eisenbrauns.

Grzimek, B. 1972. *Grzimek's Animal Life Encyclopedia, vol. 13. Mammals IV*. New York: Van Nostrand Reinhold Co.

Ingold, T. 2007. Material Against Materiality, *Archaeological Dialogues* 14/1: 1–16.

Jacobsen, Th. 1987. *The Harps that Once...: Sumerian Poetry in Translation*. New Haven and London: Yale University Press.

Jakob-Rost *et al.* 1992. L. Jacob-Rost, E. Klengel-Brandt, J. Marzahn and R.-B. Wartke, *Das vorderasiatische Museum, Staatliche Museen zu Berlin*. Mainz a.R.: von Zabern.

Knappett, C. 2014. Materiality in Archaeological Theory. In C. Smith (ed.), *Encyclopedia of Global Archaeology*: 4700-4708. New York: Springer.

Knappett, V. 2007. Materials with Materiality? *Archaeological Dialogues* 14/1: 20–23.

Landsberger, B. 1961. Einige unerkannt gebliebene oder verkannte Nomina des Akkadischen 4: anzû=(mythischer) Riesenvogel (Adler), *Wiener Zeitschrift für die Kunde des Morgenlandes* 57: 1–23.

Lambert, W.G. 1980. New Fragments of Babylonian Epics, *Archiv für Orientforschung* 27: 71–82.

Livingstone, A. 1986. *Mystical and Mythological Explanatory Works of Assyrian and Babylonian Scholars*. Oxford: Clarendon Press.

Malafouris, L. and C. Renfrew 2010 (eds), *The Cognitive Lives of Things: Recasting the Boundaries of the Mind*. Cambridge: McDonald Institute for Archaeological Research Publications.

Marchetti, N. 1996. L'aquila Anzu: nota su alcuni amuleti mesopotamici, *Vicino Oriente* 10: 105–121.

Matthews, R. 1994. Imperial Catastrophe or Local Incident? An Akkadian Hoard from Tell Brak, Syria, *Cambridge Archaeological Journal* 4/2: 290–302.

Miller, D. 2007. Stone Age or Plastic Age, *Archaeological Dialogues* 14/1: 23–27.

Moorey, P.R.S. 1999. *Ancient Mesopotamian Materials and Industries: the Archaeological Evidence*. Winona Lake: Eisenbrauns.

Moortgat, A. and U. Moortgat-Correns 1974. Archäologische Bemerkungen zu einem Schatzfund im vorsargonischen Palast in Mari, *Iraq* 36: 155–167.

Moortgat-Correns, U. 1988. Ein Kultbild Ninurtas aus neuassyrischer Zeit, *Archiv für Orientforschung* 35: 117–33.

Orthmann, W. 1975. *Der Alte Orient*, Propyläen Kunstgeschichte 14. Berlin: Propyläen Verlag.

Parrot, A. 1948. *Tello: vingt campagnes de fouilles (1877-1933)*. Paris: A. Michel.

Pongratz-Leisten, B. 2015. Imperial Allegories: Divine Agency and Monstrous Bodies in Mesopotamia's Body Description Texts. In Pongratz-Leisten and Sonik (eds): 119-141.

Pongratz-Leisten, B. and K. Sonik (eds) 2015. *The Materiality of Divine Agency*. Studies in Ancient Near Eastern Records vol. 8. Berlin and Boston: de Gruyter.

Porada, E. 1948. *Corpus of Ancient Near Eastern Seals in North American Collections. The Collection of Pierpont Morgan Library*. The Bollingen Series 14. Washington: Pantheon Books.

Renfrew, C. 2004. Towards a Theory of Material Engagement. In DeMarrais *et al.* 2004: 23–31.

Seidl, U. 1998. Das Flut-Ungeheuer abūbu, *Zeitschrift für Assyriologie und vorderasiatische Archäologie* 88: 100–113.

Sonik, K. 2015. Divine (Re-)presentation: Authoritative Images and a Pictorial Stream of Tradition in Mesopotamia. In Pongratz-Leisten and Sonik (eds): 142-193.

Tallqvist, K.L. 1938. *Akkadische Götterepitheta. Mit einem Götterverzeichnis und einer Liste der prädikativen Elemente der sumerischen Götternamen.* Studia Orientalia 7. Helsinki: Societas Orientalis Fennica.

Tilley, C. 2007. Materiality in Materials, *Archaeological Dialogues* 14/1: 16–20.

Vogelzang, M.E. 1988. *Bin sar dadme. Edition and Analysis of the Akkadian Anzu Poem.* Groningen: Styx.

Watanabe, C.E. 2002. *Animal Symbolism in Mesopotamia: a Contextual Approach.* Wiener Offene Orientalistik Bd. 1. Vienna: Institut für Orientalistik der Universität Wien.

Wengrow, D. 2004. Violence into Order: Materiality and Sacred Power in Ancient Iraq. In DeMarrais *et al.* 2004: 261–270.

Wiggermann, F.A.M. 1989. Tišpak, his Seal and the Dragon Mušhuššu. In O.M.C. Haex, H.H. Curvers and P.M.M.G. Akkermans (eds), *To the Euphrates and Beyond: Archaeological Studies in Honour of Maurits N. van Loon.* Rotterdam: Balkema.

Wiggermann, F.A.M. 1992. *Mesopotamian Protective Spirits: the Ritual Texts.* Cuneiform Monographs 1 Groningen: Styx.

Wiggermann, F.A.M. and A. Green 1994. Mischwesen (A and B). In E. Ebeling and B. Meissner (eds), *Reallexikon der Assyriologie und vorderasiatischen Archäologie,* Band 8/3-4: 222-246. Berlin: de Gruyter.

Winter, I.J. 1985. After the Battle is Over: the 'Stele of the Vultures' and the Beginning of Historical Narrative in the Ancient Near East. In H. Kessler and M.S. Simpson (eds), *Pictorial Narrative in Antiquity to the Middle Ages:* 11–32. Washington, D.C.: National Gallery of Art.

Winter, I.J. 1999. The Aesthetic Value of Lapis lazuli in Mesopotamia. In A. Caubet (ed.), *Cornaline et pierres précieuses. La Méditerranée, de l'Antiquité à l'Islam:* 45–58. Paris: Musée du Louvre.

Shining, Contrasting, Enchanting: Composite Artefacts from the Royal Tomb of Qaṭna

Elisa Roßberger
Ludwig-Maximilians-Universität München
Elisa.Rossberger@lmu.de

Abstract

In the Ancient Near East, the use of multi-coloured inlays is particularly common for rosettes and eye-like objects. The evidence from the rich jewellery inventory of the 'Royal Tomb of Qaṭna' confirms this impression for Late Bronze Age Syria. Of particular interest is a group of composite objects that resemble the anatomical shape of an eyeball and combine Baltic amber, lapis lazuli, variscite and gold. In addition to the archaeological evidence, contemporary texts help us to understand the cultural function of these artefacts' composite nature.

Keywords: Composite artefacts; jewellery; Late Bronze Age; Syria; eye-like artefact; rosette; amber

1. Introduction

Composite objects often appear to contain an element of enchantment. They captivate the viewer by combining visual and semantic qualities of distinct material elements in an act of artistic virtuosity. Much credit has been given to Alfred Gell's original conception of a 'technology of enchantment' caused by an 'enchantment of technology'.[1] According to Gell, a way to achieve this effect is the 'transformation of materials, and the ideas associated with those materials, […] into something else', basically an 'alchemy of art'.[2] When thinking about material examples for Gell's ideas in the Ancient Near Eastern record, several objects from the Royal Tomb of Qaṭna come to mind, whose striking composite nature attracts our attention.[3] While they do not mimic nature, they explicitly reference living things or essential parts thereof: eyes and flowers. Reaching back to Late Uruk Mesopotamia and recalling evidence from the Early and Middle Bronze Age Near East, we realise that it is exactly these two pictorial elements that recurrently receive multi-colour and multi-substance inlays. I will argue that their composite nature is more than just shine and contrast, but carries an element of enchantment. I will only discuss artefacts whose composite nature lies in more than just gold-framing; thus beads, pendants, seals or other artefacts consisting of ornamental stones enclosed in gold sheet or fitted with gold caps will not be included. Equally excluded was a group of toggle pins made of gold, bronze and wood.[4]

2. Composite Objects from the Royal Tomb

More than 2100 objects have been recovered from the Royal Tomb of Qaṭna, among them 1174 items defined as jewellery. More than half of that number are individual beads (607), and another large group consists of small buttons (330) originally sewn onto textiles. The composite objects discussed comprise of pendants and various kinds of large 'buttons' with fixing mechanisms at their backs.

2.1 Drop-Shaped Pendants and a Rectangular, Striped Object

The only composite objects from the Royal Tomb not resembling eyes or flowers, are a pair of drop-shaped pendants (Figure 1a)[5] and a rectangular, striped object with rows of lateral holes suggesting an original mounting on a flexible backing (Figure 1b).[6] The latter was inlaid with a dark blue, high-quality lapis lazuli

---

[1] For him, the enchantment of the viewer is one way to produce a particularly strong relation between object (index) and recipient, and it is technical, stylistic or other artistic virtuosity that causes such a reaction, often perceived as magic (Gell 1992: 43).
[2] Gell 1992: 52–53. Accordingly, the value of an artwork is not only inherent in its raw materials or production time, but also in the astonishment of the viewer that technical virtuosity can master such a radical transformation of material/substance (Gell 1992: 54).
[3] The Royal Tomb of Qaṭna, excavated in 2002 by Heike Dohmann-Pfälzner, Peter Pfälzner, Michel al-Maqdissi and their German-Syrian team, revealed a wealth of objects from an undisturbed elite burial context, mostly dating to the first half of the 14th century BCE (al-Maqdissi et al. 2003; Roßberger 2015). I thank Peter Pfälzner, University of Tübingen, for his long-lasting support and encouragement to work on this fascinating material.

[4] Roßberger 2015: 241–245, fig. 175, pls. 39–40.
[5] MSH02G-i1443 and MSH02G-i1770 (Roßberger 2015a: 150–151, fig. 113, pl. 24).
[6] MSH02G-i1011 (Roßberger 2015a: 254–255, fig. 185, pl. 42). Mineralogical analysis by J. Zöldföldi.

a) MSH02G-1443 MSH02G-i1770 b) MSH02G-i1011 |_1 cm_|

Figure 1a-b. a) Two drop-shaped pendants with variscite inlays, lapis lazuli bead between suspension loops only preserved in one case; b) rectangular object with lapis lazuli and asbestos-mineral inlays (all courtesy of the Qaṭna-Project, University of Tübingen; photos: K. Wita).

and a white flaky substance, mineralogically determined as asbestos. The pendants' light green stone inlay is not turquoise but variscite, a relatively rare phosphate mineral, more commonly found in Central Europe than in the Near East.[7] A badly weathered lapis lazuli bead between the suspension loops is only preserved in one case, but makes the ornament 'composite' according to the definition given above. Due to the severe corrosion affecting many objects recovered from the Royal Tomb's floor, the surface of the lapis lazuli bead appears brownish.[8] The shape has close parallels in Middle Bronze II and Late Bronze Age elite contexts in Egypt, the Levant and Northern Mesopotamia,[9] though the addition of a small bead in the suspension occurs only at Qaṭna. However, there is no reason to assume foreign manufacture since neither the choice of raw materials nor the techniques used differ significantly from the rest of the tomb's jewellery assemblage.

Comparisons for the striped rectangular ornament are more difficult to find. Its discovery among a group of rosettes cut from gold sheet inside the sarcophagus of the main chamber suggests that it was sewn with these items on a textile. Recalling the metope-style and rosette ornamentation of headbands worn by Iron Age ivory ladies' heads, known for instance from Tell Halaf or Nimrud, I suggested elsewhere that all of them served as decoration for a headdress.[10]

2.2 Composite Rosettes

Similarly, I reconstruct the well-known large rosette inlaid with lapis lazuli and carnelian as the central piece of a headband, attached by loops to a flexible backing (Figure 2a).[11] This complex piece of jewellery underwent a laborious conservation process that allowed us to study its composite nature.[12] Interestingly, the central round and the crescent-shaped carnelian inlays were flat and translucent with a bright red, and now powdery, substance underneath. The convex shape of the central roundel and its cylindrical drillings suggest secondary usage. A pair of smaller rosettes was made in a similar manner with thin carnelian inlays and remains of pigments found underneath (Figure 2b).[13] As a few blue pigments that still adhered to the surface suggest, the unpreserved alternate fillings were probably blue, imitating lapis lazuli, and contrasting with the orange-red of the carnelian. The large rosette was found in a group of gold sheet appliqués next to the sarcophagus in the Royal Tomb's main chamber, and I assume that they were both originally attached to the same headdress (see reconstruction in Figure 2c).

In northern Syria, rosette frontlets characterise monumental statues of female sphinxes sculptured under Hittite influence towards the end of Late Bronze and at the beginning of the Iron Age (Aleppo, Ain Dara; Figure 2d).[14] But headbands with an emblematic large circular or even rosette-shaped centre are also typical for Late Bronze Age terracotta plaques depicting females (Figure 2e). Examples from Tell Munbāqa continue the traditional motif of nude women with prominent circular frontlets going back to Early and Middle Bronze Age Syrian terracotta figurine

---

[7] The identification was carried out by a portable μ-Raman and portable XRD device by Prof. Nikai, Tokyo University (Zöldföldi 2011: 240, fn. 11).
[8] Cf. Zöldföldi 2011: 241–242.
[9] There are parallels from Thebes (Tomb of Queen Ahhotep I.), Tell al-Ajjul, Megiddo and Assur (Tomb 45). See Roßberger (2015a: 151) and Lilyquist (1993: 50, 55) for bibliographic references.
[10] See Roßberger 2015a: 254 for references. For an example, see the Metropolitan Museum New York Rogers Fund 54.117.8 (<http://www.metmuseum.org/art/collection/search/324329> last access 10 October 2016). For a discussion of rectangular forehead ornaments as female adornment see Gansell 2007.

[11] MSH02G-i1150 (Roßberger 2015a: 250–253, fig. 183, pls. 41–42).
[12] Formigli and Abbado 2011: 215–217.
[13] MSH02G-i2071 and -i2350 (Roßberger 2015a: 253–254).
[14] Roßberger 2015b; Roßberger 2015a: 235 fn. 1303.

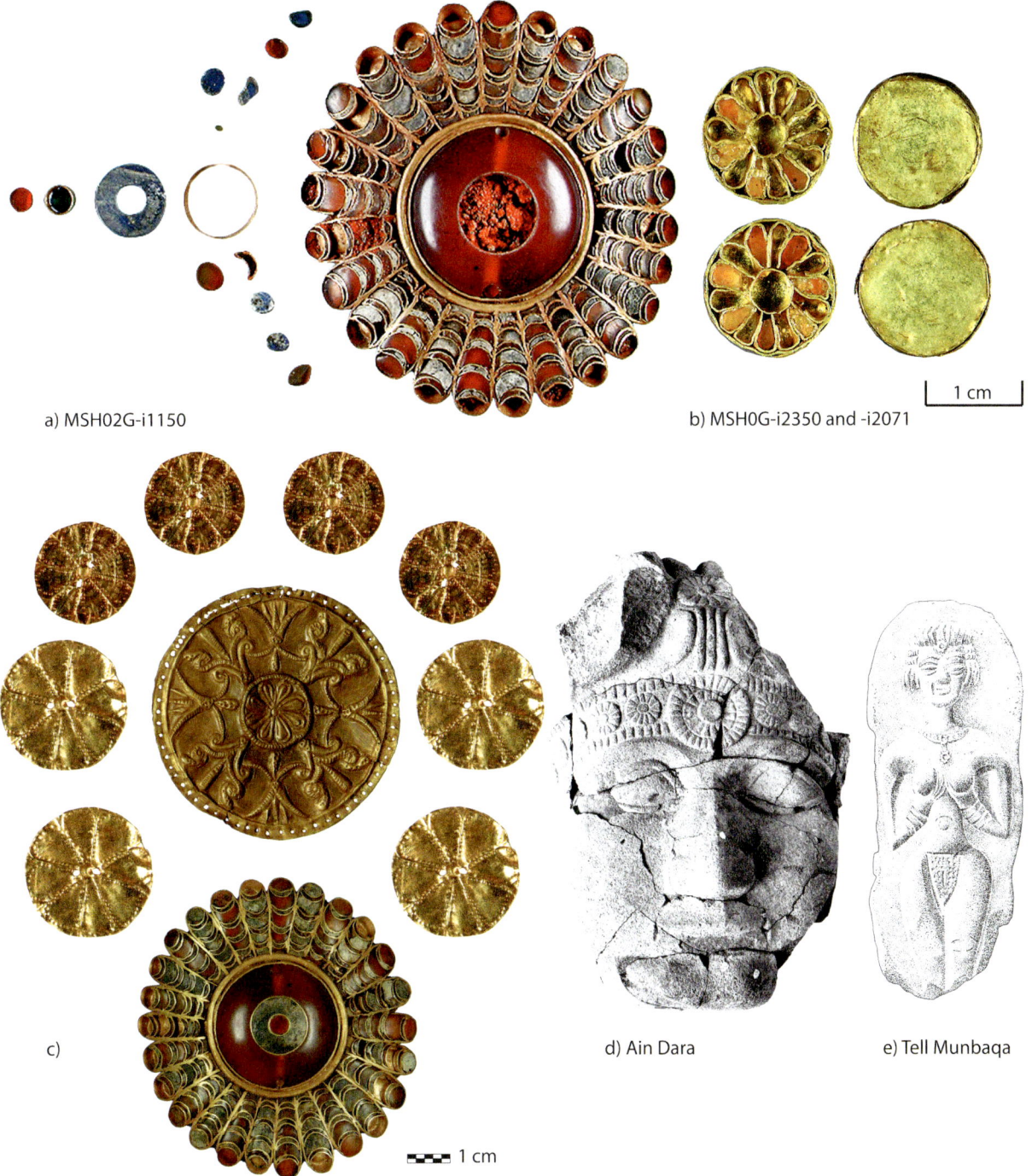

Figure 2. a) Rosette with carnelian and lapis lazuli inlays; b) small rosettes with carnelian inlays; c) reconstruction of rosette and gold-discs ensemble, all found in close distance (all courtesy of the Qaṭna-Project, University of Tübingen; photos: K. Wita; arranged by author); d) sphinx head from ʿAin Dara (after Abū-ʿAssāf, 1990, Taf. 5.56, Skulptur G 5); e) terracotta plaque from Tell Munbāqa depicting nude female with rosette frontlet (after Werner 1998, Nr. 4034).

production.[15] A headband of gold discs and a central, lapis lazuli inlaid rosette found *in situ* in the rich burial of a woman at Late Bronze Age Mari testifies to this custom; red textile fragments, found in short distance close to the head might be the remains of the original backing for these ornaments.[16] As already mentioned, such headbands continue well into the Iron Age as Syro-Levantine ivories of female heads show.[17] Neo-Assyrian kings adopted this kind of head ornamentation in their own way.

---

[15] For Late Bronze Age terracotta plaques from Tell Munbāqa see Werner 1998: pl. 156, no. 4034; Roßberger 2015b: 28–29.

[16] Jean-Marie 1999: 121, pl. 31; Nicolini 2010: 88–90, pls. 14–32.
[17] See for instance Herrmann 1992: nos. 102, 107.

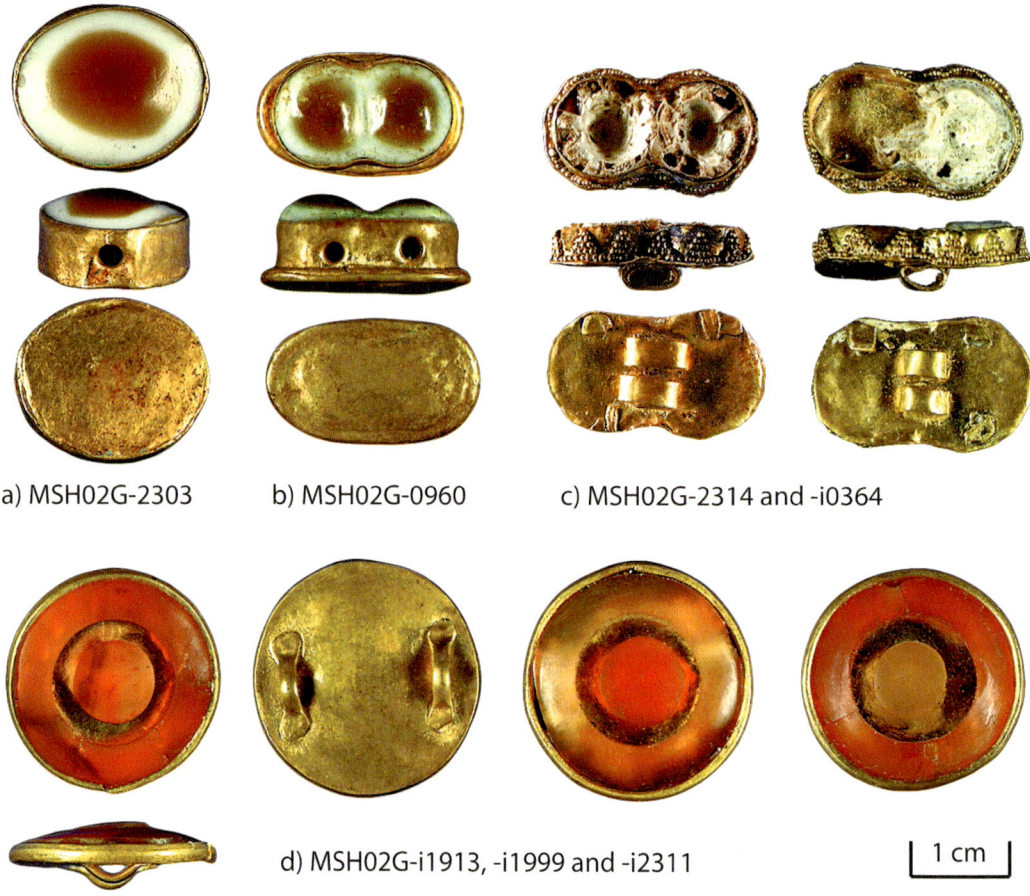

Figure 3. a) Banded agate 'eye-stone' set in gold; b) 'double-eye' banded agate set in gold;
c) two 'double-eye' jewels with glass inlays, interior filling not preserved;
d) three roundels with concentric rings of carnelian inlays, filling material not preserved
(all courtesy of the Qaṭna-Project; University of Tübingen; photos: K. Wita).

2.3 Composite Eyes

Three (but also two or four) concentric circles in divergent colours naturally evoke the impression of an eye, especially if white surrounds a black or dark centre. Round or oval-shaped, dark and white banded quartz varieties (sardonyx, onyx) serve this purpose well. They were frequently used as 'eye-stones' in early Mesopotamia, from Ur III times onwards often added by dedicatory inscriptions.[18] A gold-set example of such an eye-stone with lateral perforation comes from the Royal Tomb of Qaṭna (Figure 3a),[19] but consisting of two materials only, it does not fit our definition of a 'composite object'. The same is true for a similar 'double-eye' jewel with a banded agate inlay and two lateral perforations (Figure 3b).[20] Since details in manufacture and quality of the gold used differ for these two objects from most of the assemblage, they were probably two of the very few (Mesopotamian) imports present in the Royal Tomb.

Nevertheless, they had locally produced counter-parts with backward-loop fittings, granulated decoration and composite inlays consisting of white glass originally filled in their central depressions with a now unpreserved material (Figure 3c).[21] Again, blue pigments on the side of one button might indicate the original colour of the filling. Small, inlaid ornaments (stamp seals) with double-eye appearance occur already in Uruk times.[22] Better known are etched carnelian beads imported from the Indus valley into Early Dynastic Mesopotamia and discovered in graves of the Royal Cemetery at Ur.[23] Chronologically closer parallels for the agate version from the Royal Tomb of Qaṭna come from Middle Assyrian Assur (Tomb 45) or, made of coloured glass and without gold fittings, from Nuzi.[24] Additionally, there are many Late Bronze Age

---

[18] Clayden 2009: 41–44; Roßberger 2015a: 210–212 with references.
[19] MSH02G-i2303 (Roßberger 2015a: 210–212).
[20] MSH02G-i0960 (Roßberger 2015a: 212–213).
[21] MSH02G-i0364 and -i2314 (Roßberger 2015a: 212–214, fig. 159, pl. 37).
[22] Examples are stamp seals made of black stone with white inlays from Tell Brak (Mallowan 1947: 131, pl. XX.21) and Tepe Gawra (von Wickede 1990: 236, fig. 42.2).
[23] Roßberger 2015a: 213 fn. 1187 with references; see Zettler and Horne 1998: cat. 68 for an example.
[24] See Wartke 1994: fig. 11 for Assur and Roßberger 2015a: 213 fn. 1190–1191 for further references.

examples of simple glass beads forming concentric, eye-like circles, for instance from Nuzi or Mari.[25]

Another group of composite eye-ornaments from Qaṭna are three roundels with concentric rings consisting of thin carnelian inlays and red pigments underneath (Figure 3d).[26] Like the inlays of the rosettes, the inlays of the roundels were mounted on an unpreserved filling material. A white paste would have enhanced their eye-like appearance, though blue would be a possible option as well. The roundels form part of a group of eighteen round to oval-shaped ornamental stone jewels set in gold with back fittings for attachment on a piece of cloth.[27] I call them 'buttons', even though this term does not do justice to their original function. All 'buttons', including the double-eye buttons discussed above, lay concentrated in the area of one of the four disintegrated wooden installations in the main chamber of the Royal Tomb, making a joint mounting very likely.[28]

The most intriguing composite objects from the Royal Tomb are seven 3cm large buttons cut from amber with one or two inlays set into sockets of gold sheet (Figure 4a).[29] Slim gold stripes run through the horizontally pierced bodies, forming a loop and allowing for backward attachment (Figure 4b). What kind of object they were part of remains a mystery, since they show no signs of wear or of organic materials adhering to their surface. Typological considerations and find spots suggest that they were produced and used in pairs; in one case, two of them lay upside down only a few centimetres apart from each other in the southern part of the main chamber (Figure 4d), immediately south of the concentration of buttons referred to earlier.[30] Geologist Judit Zöldföldi determined the inlays as whitely weathered lapis lazuli and bright, greenish variscite similar to the colour combination of the drop-shaped pendants (see above 2.1) and reminiscent of an earring from Tomb 45 at Assur.[31] A gold object with concentric inlays discovered in one of the Middle Bronze Age princely tombs at Ebla could serve as comparison.[32] A closer look at the section of these objects reveals their striking resemblance with actual *eyeballs* (Figure 4c): they do not only have a size similar to human or larger animal eyes, but also does the cornea curve outwards and the strings that must have ensured attachment resemble the optic nerve connecting the eye with the brain. Unlike the typical almond-shaped and rather flat eye-inlays known from all kinds of anthropomorphic statuary from the Ancient Near East, including Middle and Late Bronze Age Syria and Qaṭna, the artisans who created these objects produced a three-dimensional version of an eye, and used highly exotic materials in contrasting and unnatural colours, all set in gold. I believe that these choices are more than just aesthetic or decorative features.

3. Discussion

3.1 Ancient Near Eastern Traditions of Composite Eyes and Rosettes

Above, I referred to a rosette headband from Mari whose centre piece had a lapis lazuli inlay. The tradition of forming rosette decorations of multi-coloured materials in the Ancient Near East dates back much further in time, with white-black-and-red rosette wall-pegs and mosaic inlays for stone vessels and other artefacts made of perishable material from Late Uruk/Jemdet Nasr Mesopotamia.[33] Interestingly, besides rosettes, *eyes* had been inlayed in these early objects, for instance in the well-known spouted vessels from the 'Sammelfund' at Uruk;[34] a mosaic from the same context combines the two motifs with three concentric circles ('eye') as a rosette's centre piece,[35] similar to the design of the large rosette from Qaṭna, and many rosettes known from Middle- and Neo-Assyrian as well as Achaemenid reliefs and wall-paintings.[36] Early and Middle Bronze Age examples of inlaid rosette-decorated jewellery are well attested in Mesopotamia and beyond.[37] Browsing through the rich assemblage from the Neo-Assyrian Queens' Tombs at Nimrud offers a last piece of evidence that – through the ages – eyes, rosettes, or a combination of both, were the ornaments most likely to receive multi-coloured inlays in Ancient Near Eastern jewellery production.[38] Of particular interest is the lower part of the famous crown of Queen Hama consisting of a skillful arrangement of golden rosettes and 63 oval balls with inlaid centres, conventionally interpreted as 'poppy capsules'.[39] Considering the similarity of their shapes with that

---

[25] For references see Roßberger 2015a: 210–211 fn. 1168.
[26] MSH02G-i1913, -i1999 and -i2311 (Roßberger 2015a: 206–210).
[27] Roßberger 2015a: 206–210.
[28] Roßberger 2015a: fig. 152.
[29] MSH02G-i1376 (only gold fitting preserved), -i0991, -i1472, -i1949, -i1950, -i2106, -i2174 (Roßberger 2015a: 213–216, figs. 161–164, pls. 37–38).
[30] Roßberger 2015a: fig. 160.
[31] Wartke 1994: fig. 14.
[32] Matthiae 1979: 176, fig. 79.
[33] Heinrich 1936: pls. 26–28a, 32; Aruz 2003: fig. 45a–c.
[34] Heinrich 1936: pls. 26, 28b.
[35] Heinrich 1936: pl. 32.
[36] Cf. Musche 1994: fig. 1.
[37] For examples from the Royal Cemetery at Ur, Tello and Acemhöyük see Roßberger 2015a: 250–251 with references. Examples of the many composite rosettes from Ur can be found in Zettler and Horne 1998: cat. 29, 34, 54-55, 94. Eyes and rosettes are also the most frequent motifs inlaid in gaming boards from the Royal Cemetery. From Egypt, we have iconographic and archaeological evidence for headbands decorated with inlaid rosettes from the tombs of high-ranking females from the Old Kingdom onwards (see Roßberger 2015a: 234–235, 250–253 and Wilkinson 1971: 37–43, 70–71, 114–115 with references).
[38] See now Hussein et al. 2016: e.g. pls. 50–54, 67c, 70, 72–75b, 160. Pl. 54b depicts a gold rosette with a concentrically banded agate centre-piece; pl. 160g–h gold rosettes with eye-stone centre-pieces (original stringing not recorded).
[39] ND 1989.309, found on the head of a female skeleton in Tomb 3, coffin 2 (Hussein et al. 2016: 123, pls. 129–132; Collon 2008: 105–106).

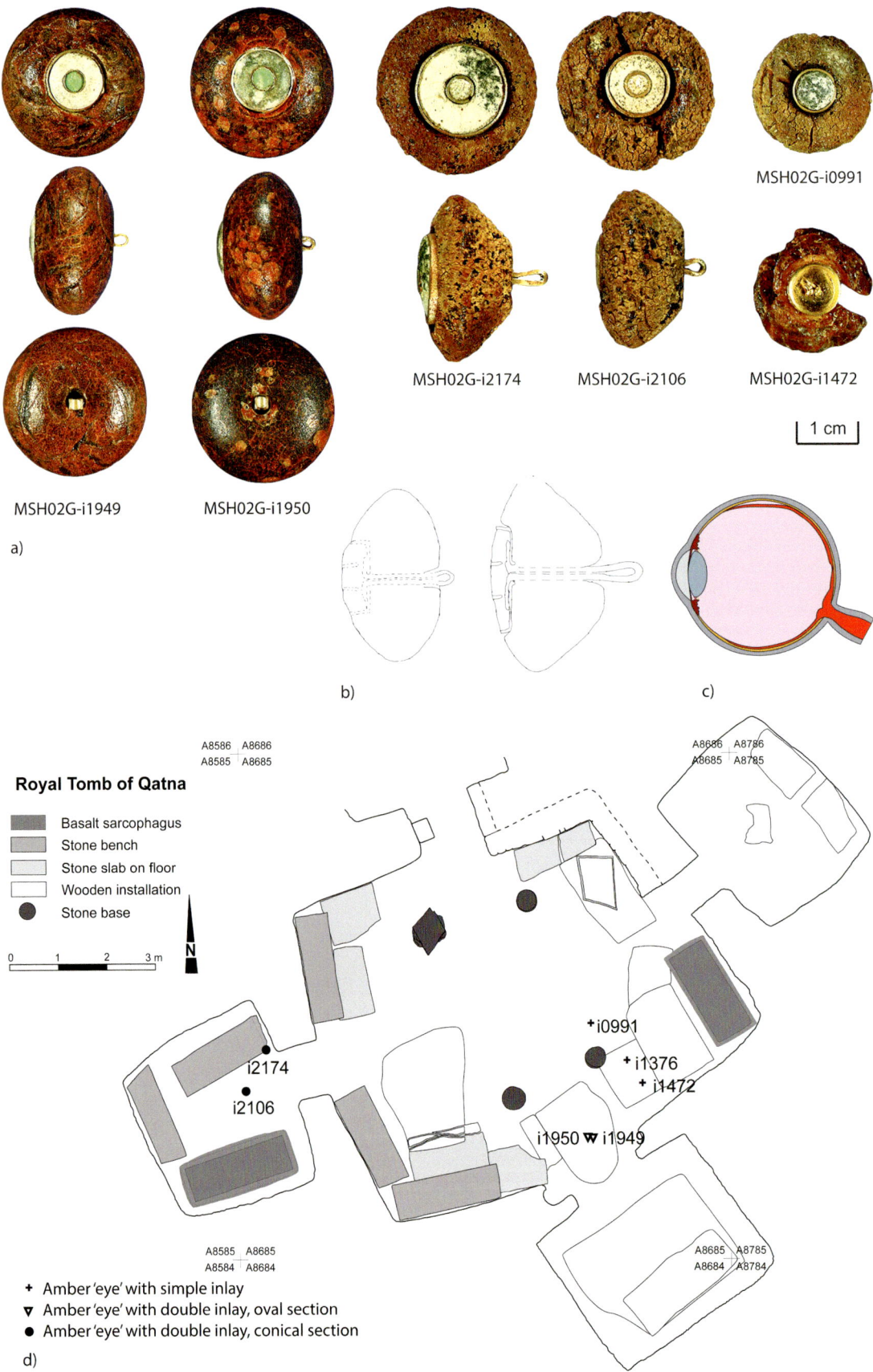

Figure 4. a) Eye-like artefacts consisting of amber, lapis lazuli, variscite and gold strips; b) cross sections of two of the artefacts (both courtesy of the Qaṭna-Project, University of Tübingen; photos: K. Wita); c) schematic diagram of the human eye (Wikimedia commons; by Erin Silversmith from an original by en:User:Delta G); d) find spots of eye-like amber artefacts at the Royal Tomb of Qatna (author, based on plan drawing by A. Bianchi et al., Qaṭna Project Tübingen).

of the large amber 'eyes' from Qatna (2.3), I suggest to interpret the crown's decoration as a combination of rosettes and eye(balls) following an age-old tradition for Mesopotamian headdresses.[40]

3.2 Eyes of Anthropomorphic Statuary: Channeling the Divine Gaze

It comes as no surprise, that Ancient Near Eastern anthropomorphic statues and figurines had composite eyes made of materials different from the rest of their bodies, no matter if they consisted of stone, metal or wood, and regardless if they depicted deities, kings or other human worshippers.[41] If we take the numerous Early Dynastic worshipping statuettes as an example, we observe that their eyes were rarely carved and/or painted. Instead, we find inlays even in very small or badly made examples, and even if the colour of the stone used for the statue matches that of the eye inlay very closely, for instance shell inlays in white lime stone or alabaster statuettes.[42] While the sclera (the 'white of the eye') of Mesopotamian stone statues was often made of shell, for Syrian metal or wood statues, limestone and steatite inlays were frequently employed, for instance in royal statuary known from Hazor or Qatna.[43] Very rarely do eye inlays occur in relief sculpture.[44]

In my opinion, these choices do not only aim at naturalistic visual effects, but also at a divergence of substance. This becomes most evident when the colours of the background and inlay material are similar. The small mask of a woman discovered at the Lower City Palace at Qatna illustrates this point: while the whole mask consists of ivory, the eyes are inlaid with shell (sclera) and translucent gypsum (iris) – all similar in colour but very different in nature.[45] The practice of applying eyes – mostly in contrasting materials – to anthropomorphise artefacts reaches back to Pre-Pottery Neolithic times with white shell (and occasionally black flint) inlaid modelled skulls from the Levant and Anatolia,[46] and a male stone-statue from the Urfa region.[47] The effect works even in non-anthropomorphic forms: cowrie shells or circular beads set into early second millennium brick-like lumps of clay,[48] and multi-coloured stone rings inserted into shells[49] or stone vessels[50] from Early Dynastic times in the Diyala region, result in 'facial features' for inanimate things and might have caused the viewer to react to them as if they were living beings. Such efforts support Silvana di Paolo's point that an artefact's composite nature is crucial for its agency and its connecting function between material and mental spheres.[51]

Ritually 'opening the eyes' was part of the Babylonian version of the mouth-opening ritual (*mīs pî*) performed on Mesopotamian statues from the late 3rd millennium onwards.[52] A recent Assyriological study by Ainsley Dicks on words of vision in literary cuneiform sources stresses the crucial importance of eye contact in the encounter between human and divine, and the semantic difference between an outward-going, powerful gaze attributed to gods, and an inward-moving sight typical for lower-ranking human beings.[53] Consequently, Gudea orders his statue: 'Oh statue, your eye is that of Ningirsu', thereby asking it for enduring attention directed towards his god.[54] I would add that eyes are the bodily orifice that allow the encounter between god and human worshipper to happen; they are the organs that open a channel between the interior and the exterior of the body, and eventually between the human and divine worlds.[55] Besides, intellectual capacities and wisdom were not considered to reside

---

[40] Unfortunately, only bitumen as adhesive material for the original inlays is preserved in some of the 'eyeballs' centres (Collon 2008: 106).
[41] Renger 1980–83: 310–311.
[42] According to the Diyala Archaeological Database (<http://diyala2.uchicago.edu> last access 10 October 2016), most eye-inlays of Diyala worshipper statuettes consist of shell. Go to Kh. IV 288, Kh. IV 289a, Kh. IV 323, Kh. VIII 115 or Kh. VIII 202 for examples of small (heads up to 5cm tall) and simply made statuettes that still feature prominent shell and bitumen inlaid eyes.
[43] The eyes of a small bronze-statuette from Hazor consist of silver (framing), white steatite (sclera) and black stone (pupils) (Ornan 2012: 447, fig. 3). The inlays of the basalt statues flanking the entrance of the Royal Tomb of Qatna had steatite inlays (sclera), the iris-inlays were lacking (Katalog Qatna 2009: 204). The eyes of the famous Idrimi magnesite-statue (whitish stone) from Alalakh, on the other hand, were inlaid with glass (iris/lens; British Museum, BM 130738).
[44] Few Iron Age basalt orthostats from Tell Halaf have inlaid eyes (Orthmann 1971: pl. 12 a–c).
[45] Luciani 2006: fig 11.
[46] For an overview on PPNB-plastered skulls, most of them with inlaid eyes, see Bonogofsky 2006: 15. A plastered skull from Yiftahel held white shell and black flint as eye inlays (Slon et al. 2014: 3, fig. 4).
[47] For a colour photograph of the PPNA stone-statue from Yeni Mahalle see Katalog Anatolien 2007: 71.
[48] Two cube-shaped clay objects (7–8cm) from early second millennium Isin (IB-1177 and 1178) feature white circular shell inlays (= eyes), a lengthy bead of lapis lazuli (= nose) and a small carnelian ring (= mouth) respectively (Hrouda 1978, fig. 3a–b). A brick-like (l. ~10cm) clay object from a Isin-Larsa period domestic quarter (EM, Room 5 'house chapel') received anthropomorphic looks through inserted cowrie shells and a pinched nose (U. 7587; see Ur online: <http://www.ur-online.org/subject/6880/> last access 10 October 2016). A similar object (U.17428; 13.8 x 10.4cm; AH House XIII) had a row of four cowrie shells inserted in one side (Reade 2002: fig. 2; http://www.ur-online.org/subject/18877/ last access 10 October 2016). U.18727 with one carnelian bead and two cowries might have been a similar object (<http://www.ur-online.org/subject/20113/> last access 10 October 2016; without illustration).
[49] This kind of artefacts occurs in Early Dynastic III contexts at Khafajeh (Kh. II 4, Kh. IX 86, Kh. II 63 (<http://diyala2.uchicago.edu/> last access 10 October 2016).
[50] See a pair of inlaid conical bowls from Early Dynastic I Khafajeh (Sin Temple IV; Kh. VI 373 and 374 (<http://diyala2.uchicago.edu/> last access 10 October 2016).
[51] On this aspect, see the article by S. Di Paolo in this volume, especially p. 15.
[52] Berlejung 1998: 244, 249.
[53] Dicks 2012; A. Dicks (Hawthorn) presented her results in a lecture at Munich University on 14.07.2014.
[54] Statue B, vii 58–59 following Edzard 1997: 36.
[55] Similarly, Edzard (1997: 36), citing P. Steinkeller: "Ningirsu could 'see' through the statue which was to serve as a 'channel' between Gudea and the god". Both, D. Freedberg (1989: 85–86) and I. Winter (2000) pointed to the centrality of viewing and eye contact between cult image and devotee (*darśan*) in Indian Hinduistic and Buddhist worship, highlighting the impact of the deity's presence.

in the human brain, but rather to be absorbed through ears and eyes.[56] Returning to Late Bronze Age Syria, the act of unveiling Dagan's face and thereby establishing sight between cult image and devotees was of high ritual significance as documented at Emar.[57]

3.3 Composite Jewellery Items in Contemporary Textual Sources

We are lucky that we do not only have archaeological but also textual evidence for the riches of Late Bronze Qaṭna. The long inventories of the palace and city goddess Bēlet-ekalli mention three jewellery items called *īn immeri* (IGI.UDU) 'eye of a sheep'.[58] In one instance this item consists, quite suitably for the archaeological record, of *pappardilû*-stone, meaning banded agate.[59] In another case the object is equally made of *pappardilû*, but with additional inlays in lapis lazuli and *dušû*-stone, which is probably to be identified with a bright green stone,[60] set in gold. Besides the replacement of agate with amber, this artefact must have looked similar to the large eyeballs found in the Royal Tomb (see above 2.3, Fig. 4a). Besides, Mesopotamian ritual texts know of variations of animal eyes like 'fish eye', 'snake eye', 'pig eye' and 'bird eye' made of *ḫulālu*- or *muššarru*-stone and often set in gold.[61] Their association with kingship is attested in the so-called 'necklace of Hammurapi' consisting of 14 stones with a 'fish eye-stone' (IGI.KU$_6$) listed at the very beginning.[62] An Amarna letter enumerating gifts from the Babylonian to the Egyptian king starts with a pair of 'eyestones' made of *pappardilû*- and *muššarru*-stones.[63] The practice of wearing necklaces with eyestones is usually explained with a presumed apotropaic effect of these objects against the 'evil eye', a concept that is actually only rarely attested in cuneiform sources.[64]

However, I would suggest that the textually well documented capability of eyes and vision to connect the human and the divine (see above), leads to a better understanding of the eyestones' function. The (animal) eye might have functioned as an icon, allowing the beneficent divine gaze to enter a human being who passively perceived its emanating power. Two other artefact categories are notable from the Qaṭna inventories for their composite set-up that equally matches the archaeological record. The first are so-called *ṣiṣṣatu*-ornaments, that is blossom-shaped items made out of gold with multiple inlays in two or three different materials (lapis lazuli, carnelian, *dušû/marḫaše*-stone) – reminiscent of the inlaid rosettes known from the Royal Tomb (see above 2.2, Fig. 2a-b). In two instances, the item is further described as 'made in the manner of the land Tukriš' (*ša qāti* ^{kur}*tukriš*) thus pointing to the foreign nature of the item or its manufacturing technique.[65]

Another type of inlaid artefacts from the Qaṭna-inventories are 'sun-discs' (Akkadian *šamšatu*), a common form of circular pendant often dedicated to gods, especially during the Old Babylonian period.[66] Mostly made of gold, some sun-discs had additional ornaments in lapis lazuli and *dušû* attached to them. While a golden sun-disc discovered at the Royal Tomb of Qaṭna had no inlays,[67] we have good archaeological evidence for inlaid examples from Middle Bronze Age Ebla and Byblos,[68] and in two simpler versions from Late Bronze Age Alalakh and Tell Bazi.[69] These examples are decorated with inlaid circular and crescent shaped forms that are explicitly mentioned in the texts as decorations for these objects (*alu, kussu, puku, gullatu* und *sînu*-'moon-crescent'-ornaments).[70] The colours of the stones mentioned in the texts correspond with the preferences evident in the finds from the Royal Tomb: blue, red and green, preferably made of stones but occasionally replaced by vitreous materials (see the double eye buttons in 2.3, Fig. 3b-c). The sun-disc medallion from Byblos preserved an inlay consisting of a composite 'eye-stone' made of a white outer and an unpreserved, probably darker, inner part.

4. Conclusion

To summarise the main characteristics of composite objects from the Royal Tomb of Qaṭna. The two drop-shaped pendants (2.1), as well as all the jewellery items with backward mechanisms for attachment (2.2–2.3) have a clear presentation side directed towards the viewer. All of them are set in gold with inlays in contrasting colours made of stones or glass. The pendants and the rectangular object aside (2.1), they resemble rosettes or eyes. This observation matches the contemporary textual record from Qaṭna, that mentions flower- and (animal) eye-shaped ornaments with multiple inlays (3.3).

---

[56] Steinert 2012: 386 fn. 7.
[57] Fleming 2015: 205.
[58] Bottéro 1949: Inv. I. 41, 55, 60, 86, 123; Inv. III r.16; Roßberger 2015a: 344.
[59] Schuster-Brandis 2008: 403–404.
[60] Schuster-Brandis 2008: 407; Roßberger 2015a: 355.
[61] Schuster-Brandis 2008: 418. Eyestones made of banded agate or glass, sometimes set in gold and often inscribed, were among the most frequent Mesopotamian royal votive gifts especially in the late third and second millennium.
[62] Schuster-Brandis 2008: 167–168.
[63] EA 13, 3–4.
[64] For the very limited evidence on beliefs in an 'evil eye' from Sumerian and Akkadian sources see Thomsen (1992: 28) and Geller (2003: 115-134).

[65] Roßberger 2015a: 352.
[66] See now Arkhipov 2012: 93–96 with references.
[67] MSH02G-i0829 (Roßberger 2015a: 151–152).
[68] For the piece from Ebla see Matthiae *et al.* (1995: cat. 400), and for Byblos, 'Jarre Montet', Montet (1928/29: pl. LXIII.411).
[69] For the piece from Alalakh see Woolley 1955: fig. 70, and for Tell Bazi Otto 2006: fig. 67.7.
[70] Roßberger 2015a: 342, tables 102, 358–361.

I argue that the rosette-like pieces originally adorned headbands, and were probably worn by women. Reconstructing the original attachment of the eye-like buttons is impossible from the archaeological record but I find it highly likely that their setup represents eyeballs. We might go further and suggest that the large rosette's combination of a concentric and markedly convex centre surrounded by radiating petals symbolises an eye inscribed in a flower, allowing beneficent divine emission to enter the wearer's forehead. A tripartite sub-division of a circle matches an eye's visible structure – and I would be confident that this was an icon commonly used and understood in Late Bronze Age Syria, as it was in Greater Mesopotamia from at least the 3rd millennium onwards. Mimicking natural colouration was obviously not important – at least not in the case of the large 'eyeballs' from the Royal Tomb made of amber, lapis lazuli and variscite. Here, we might encounter an intentional choice of exotics, raw materials with unusual characteristics brought from afar – probably still resonating in designations like *marhaše* 'stone from the land Marhaši', or *ša qāti Tukriš* 'made in the manner of the land Tukriš'.

While the exact qualities associated with the materials used elude us, we may conclude, that their colourful mixing and concentric arrangement was an appropriate material solution for referencing the transition between interior and exterior of a human and/or divine body, an artistic implementation of what Alfred Gell described as 'homunculus effect'.[71] The artisans who made these objects found a way to materialise the channel between animate and inanimate worlds and – most likely – to enchant their viewers.

References

al-Maqdissi, M., H. Dohmann-Pfälzner, P. Pfälzner and A. Suleiman 2003. Das königliche Hypogäum von Qaṭna. Bericht über die syrisch-deutsche Ausgrabung im November-Dezember 2002, *Mitteilungen der Deutschen Orient-Gesellschaft* 135, 2003: 189–218.

Abū-ᶜAssāf, A. 1990. *Der Tempel von ᶜAin Dārā*, Damaszener Forschungen 3. Mainz am Rhein: von Zabern.

Arkhipov, I. 2012. *Le vocabulaire de la métallurgie et la nomenclature des objets en métal dans les textes de Mari*, Matériaux pour le Dictionnaire de Babylonien de Paris 3, ARM 32. Louvain and Paris: Peeters.

Aruz, J. 2003. *Art of the First Cities. The Third Millennium B.C. from the Mediterranean to the Indus*. New York and New Haven: Metropolitan Museum of Art and Yale University Press.

Berlejung, A. 1998. *Die Theologie der Bilder. Das Kultbild in Mesopotamien und die alttestamentliche Bilderpolemik unter besonderer Berücksichtigung der Herstellung und Einweihung der Statuen*, Orbis Biblicus et Orientalis 62. Fribourg and Göttingen: Universitätsverlag and Vandenhoeck & Ruprecht.

Bonogofsky, M. 2006. Complexity in Context: Plain, Painted and Modelled Skulls from the Neolithic Middle East. In M. Bonogofsky (ed.), *Skull Collection, Modification and Decoration*: 15–28. BAR International Series 1539. Oxford: Archaeopress.

Bottéro, J. 1949. Les inventaires de Qatna, *Revue d'assyriologie et d'archéologie orientale* 43: 1–40, 137–213.

Braun-Holzinger, E.A. 1977. *Frühdynastische Beterstatuetten*, Abhandlungen der Deutschen Orient-Gesellschaft 19. Berlin: Gebr. Mann Verlag.

Clayden, T. 2009. Eye-stones, *Zeitschrift für Orient-Archäologie* 2: 36–87.

Collon, D. 2008. Nimrud Treasures: Panel Discussion. In J.E. Curtis, H. McCall, D. Collon and L. al.-Gailani Werr (eds), *New Light on Nimrud. Proceedings of the Nimrud Conference 11th–13th March 2002*: 105–118. London: British Institute for the Study of Iraq and The British Museum.

Dicks, A.A. 2012. *Catching the Eyes of the Gods: The Gaze in Mesopotamian Literature*. Ph.D. Dissertation. New Haven: Yale University.

Edzard, D.O. 1997. *Gudea and His Dynasty*, The Royal Inscriptions of Mesopotamia. Early Periods 3/1. Toronto and Buffalo: University of Toronto Press.

Fleming, D. 2015. Seeing and Socializing with Dagan at Emar's *zukru* Festival. In B. Pongratz-Leisten and K. Sonik (eds), *The Materiality of Divine Agency*. Studies in Near Eastern Records 8: 197–210. Berlin: de Gruyter.

Formigli, E. and M. Abbado 2011. Die technologische Analyse der Goldobjekte aus der Königsgruft. In P. Pfälzner (ed.), *Interdisziplinäre Studien zur Königsgruft von Qatna*. Qaṭna Studien 1: 191–235. Wiesbaden: Harrassowitz.

Freedberg, D. 1989. *The Power of Images. Studies in the History and Theory of Response*. Chicago: University Press.

Gansell, A.R. 2007. From Mesopotamia to Modern Syria. Ethnoarchaeological Perspectives on Female Adornment during Rites of Passage. In J. Cheng and M.H. Feldman (eds), *Ancient Near Eastern Art in Context. Studies in Honor of Irene J. Winter by her Students*: 449–483. Leiden and Boston: Brill.

Gell, A. 1992. The Technology of Enchantment and the Enchantment of Technology. In J. Coote and A. Shelton (eds), *Anthropology, Art, and Aesthetics*: 40–63. Oxford: Clarendon Press.

Gell, A. 1998. *Art and Agency: An Anthropological Theory*. Oxford: Clarendon.

Geller, M.J. 2003. Paranoia, the Evil Eye, and the Face of Evil. In W. Sallaberger, K. Volk and A. Zgoll (eds), *Literatur, Politik und Recht in Mesopotamien. Festschrift für Claus Wilcke*, Orientalia biblica et christiana 14: 115–134. Wiesbaden: Harrassowitz.

---
[71] Gell 1998: 133.

Heinrich, E. 1936. *Kleinfunde aus den archaischen Tempelschichten in Uruk*, Ausgrabungen der Deutschen Forschungsgemeinschaft in Uruk-Warka 1. Berlin: Harrassowitz.

Herrmann, G. 1992. *The Small Collections from Fort Shalmaneser*, Ivories from Nimrud V London: The British School of Archaeology in Iraq.

Hrouda, B. 1978. Zu einem Terrakottarelief aus Isin, *Zeitschrift für Assyrologie und Vorderasiatische Archäologie* 68: 280–283.

Hussein, M.M., M. Altaweel and M. Gibson 2016. *Nimrud. The Queens' Tombs*, Oriental Institute Miscellaneous Publications. Baghdad and Chicago: Iraqi State Board of Antiquities and Heritage and The Oriental Institute of Chicago.

Jean-Marie, M. 1999. *Tombes et nécropoles de Mari*, Mission archéologique de Mari 5. Bibliothèque archéologique et historique 153. Beyrouth: Institut français d'archéologie du Proche-Orient.

Katalog Anatolien 2007. *Vor 12.000 Jahren in Anatolien. Die ältesten Monumente der Menschheit, Begleitbuch zur großen Landesausstellung Baden-Württemberg 2007 im Badischen Landesmuseum Schloss Karlsruhe*. Stuttgart: Theiss.

Katalog Qatna 2009. *Schätze des Alten Syrien. Die Entdeckung des Königreichs Qatna, Begleitbuch zur Großen Landesausstellung Baden-Württemberg im Württembergischen Landesmuseum Stuttgart*. Stuttgart: Theiss.

Lilyquist, C. 1993. Granulation and Glass. Chronological and Stylistic Investigations at Selected Sites, ca. 2500–1400 B.C.E., *Bulletin of the American School of Oriental Research* 290–291: 29–94.

Luciani, M. 2006. Palatial Workshops at Qatna?, *Baghdader Mitteilungen* 37: 403–430.

Mallowan, M.E.L. 1947. Excavations at Tell Brak and Chagar Bazar, *Iraq* 9: 1–259.

Matthiae, P. 1979. Scavi a Tell Mardikh-Ebla, 1978: rapporto sommario, *Studi Eblaiti* I, 9-12, 129–184.

Matthiae, P., F. Pinnock and G. Scandone Matthiae (eds), Ebla. *Alle origini della civiltà urbana. Trent'anni di scavi in Siria dell'Università di Roma 'La Sapienza'*. Milano: Electa.

Maxwell-Hyslop, K.R. 1971. *Western Asiatic Jewellery c. 3000-612 BC*. London: Methuen & Co.

Montet, P. 1928/29. *Byblos et l'Egypte. Quatre campagnes de fouilles à Gebeil*, Bibliothèque Archéologique et Historique 11. Paris: Librairie orientaliste Paul Geuthner.

Musche, B. 1994. Zur altorientalischen Rosette. Ihr botanisches Vorbild und dessen pharmazeutische Verwertung, *Mesopotamia* 29: 49–71.

Nicolini, G. 2010. *Les ors de Mari*. Beyrouth: Institut Français du Proche-Orient.

Ornan, T. 2012. The Role of Gold in Royal Representation. The Case of a Bronze Statue from Hazor. In R. Matthews and J. Curtis (eds), *Proceedings of the 7th International Congress on the Archaeology of the Ancient Near East, London, April 12th-16th 2010*, Vol. 2: 445–458. Wiesbaden: Harrassowitz.

Orthmann, W. 1971. *Untersuchungen zur späthethitischen Kunst*, Saarbrücker Beiträge zur Altertumskunde 8. Bonn: Rudolf Habelt Verlag.

Otto, A. 2006. *Alltag und Gesellschaft zur Spätbronzezeit: eine Fallstudie aus Tall Bazi (Syrien)*, Subartu 19. Turnhout: Brepols.

Reade, J.E. 2002. Unfired Clay, Models, and 'Sculptors' Models' in the British Museum, *Archiv für Orientforschung* 48–49: 147–167.

Renger, J. 1980-1983. Kultbild. A. Philologisch. In D.O. Edzard (ed.), *Reallexikon der Assyriologie und Vorderasiatischen Archäologie* 6: 314–319. Berlin and New York: de Gruyter.

Roßberger, E. 2015a. *Schmuck für Lebende und Tote. Form und Funktion des Schmuckinventars der Königsgruft von Qatna in seinem soziokulturellen Umfeld*, Qaṭna-Studien 4. Wiesbaden: Harrassowitz.

Roßberger, E. 2015b. Schmuck für Lebende und Tote. Blüten und Pflanzen im Schmuckinventar der Königsgruft von Qaṭna, *Alter Orient aktuell* 13: 25–31.

Schuster-Brandis, A. 2008. *Steine als Schutz- und Heilmittel. Untersuchung zu ihrer Verwendung in der Beschwörungskunst Mesopotamiens im 1. Jt. vor Chr.*, Alter Orient und Altes Testament 46. Münster: Ugarit-Verlag.

Slon, V., R. Sarig, I. Hershkovitz, H. Khalaily, and I. Milevski 2014. The Plastered Skulls from the Pre-Pottery Neolithic B site of Yiftahel (Israel). A Computed Tomography-Based Analysis. *PLoS ONE* 9, 2: 1–9.

Steinert, U. 2012. *Aspekte des Menschseins im Alten Mesopotamien. Eine Studie zu Person und Identität im 2. und 1. Jt. v. Chr.*, Cuneiform Monographs 4. Leiden: Brill.

Thomsen, M.L. 1992. The Evil Eye in Mesopotamia, *Journal of Near Eastern Studies* 51: 19–32.

von Wickede, A. 1990. *Prähistorische Stempelglyptik in Vorderasien*, Münchner Vorderasiatische Studien 6. München: Profil Verlag.

Wartke, R.-B. 1994. Die mittelassyrische Gruft 45 aus Assur. Fundgeschichte, Beigaben und Rekonstruktion im Berliner Vorderasiatischen Museum, *Antike Welt* 25, 3: 237–251.

Werner, P. 1998. Terrakottareliefs. In R.M. Czichon and P. Werner (eds), *Tall Munbaqa - Ekalte I. Die bronzezeitlichen Kleinfunde*, Wissenschaftliche Veröffentlichungen der Deutschen Orient-Gesellschaft 97: 307–332. Saarbrücken: Saarbrücker Druckerei und Verlag.

Wilkinson, A. 1971. *Ancient Egyptian Jewellery*. London: Methuen.

Winter, I.J. 2000. Opening the Eyes and Opening the Mouth: The Utility of Comparing Images of Worship in India and the Ancient Near East. In M.W. Meister (ed.), *Ethnography and Personhood. Notes from the*

Field: 129–162. Jaipur and New Delhi: Rawat Publications.

Woolley, C.L. 1955. *Alalakh. An Account of the Excavations at Tell Atchana in the Hatay, 1937–1949*. Oxford: The Society of Antiquaries.

Zettler, R.L. and L. Horne (eds) 1998. *Treasures from the Royal Tombs of Ur*. Philadelphia: University of Pennsylvania Museum of Archaeology and Anthropology.

Zöldföldi, J. 2011. Gemstones at Qaṭna Royal Tomb: Preliminary report. In P. Pfälzner (ed.), *Interdisziplinäre Studien zur Königsgruft von Qaṭna*, Qaṭna Studien 1: 235–248. Wiesbaden: Harrassowitz.

Entangled Relations over Geographical and Gendered Space: Multi-Component Personal Ornaments at Hasanlu

Megan Cifarelli
Professor of Art History, Manhattanville College (NY)
megan.cifarelli@mville.edu

Abstract

Hasanlu, in Northwestern Iran, is best known for its catastrophic destruction ca. 800 BC, likely at the hands of the Urartian army. Excavations of the site revealed more than 100 burials from the period leading up to the destruction, Hasanlu Period IVb (1050–800 BC). Among these burials are five adult women decorated with multicomponent personal ornaments consisting of repurposed copper alloy or iron armour scales with attached garment pins, stone, shell and composite beads, and copper alloy tubes of various lengths. If worn on the body during life, these objects would have been both visually and aurally conspicuous. Bead and tube elements are typical of the material culture of Hasanlu, used in mortuary jewellery from the Middle Bronze Age forward. The armour scales, however, are found only in these few female burials at Hasanlu. In the broader ancient Near East, scale armour is associated with representations of male bodies in military contexts, and is found archaeologically in military, palatial, cultic and mortuary contexts. These particular scales are characteristic of regions to Hasanlu's north (the South Caucasus) and east (the Caspian littoral). This paper proposes that the creation of composite objects from these parts—fragments of masculine armour, components of personal adornment, and sound making tubes—entangled people and things across gendered and geographical boundaries.

Keywords: Hasanlu, Mortuary Archaeology, Archaeology of Gender, Entanglement, Personal Adornment, Dress, Militarisation; Armour Scales

Introduction

Hasanlu is a site in the Ushnu Solduz valley in Northwestern Iran (Figure 1) that was excavated by a joint expedition of the University of Pennsylvania Museum, the Metropolitan Museum of Art, and the Iranian Antiquities department, between 1956–1977 under the leadership of Robert H. Dyson. The site has a relatively large horizontal expanse and a long occupation sequence, beginning in the Prepottery Neolithic and stretching into the Medieval period. It consists of a high mound, or citadel, with a core of monumental buildings appearing in the Late Bronze Age (Hasanlu V, 1450–1250 BC) and developing continuously through the destruction at the end of Hasanlu IVb (1050–800 BC) (Figure 2). Surrounding the citadel is the lower mound, the site of burials from the Early Bronze Age forward.[1]

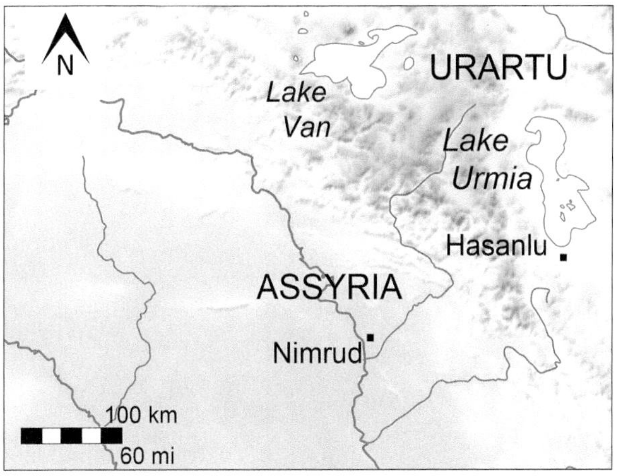

Figure 1 Map showing location of Hasanlu (base map Wikimedia Commons).

This paper deals with durable goods relating to dress that come from the burials of Period IVb, the period in which Hasanlu reaches its zenith in terms of wealth, and one which ends with its dramatic, total destruction and abandonment at the hands of the Urartian army in 800 BCE. In particular, this paper examines a group of multi-component ornaments that were found on the bodies in the best furnished women's burials of Period IVb.

While the material culture of Hasanlu displays continuous development from the Middle Bronze Age through the destruction, there is archaeological evidence from both the citadel and the burials for a heightening of social hierarchy in Period IVb, with increasing restriction of access to buildings on the citadel and the goods therein, as well as an

---
[1] For a thorough analysis of the archaeology of Hasanlu from the Middle Bronze through early Iron Age, see Danti 2013.

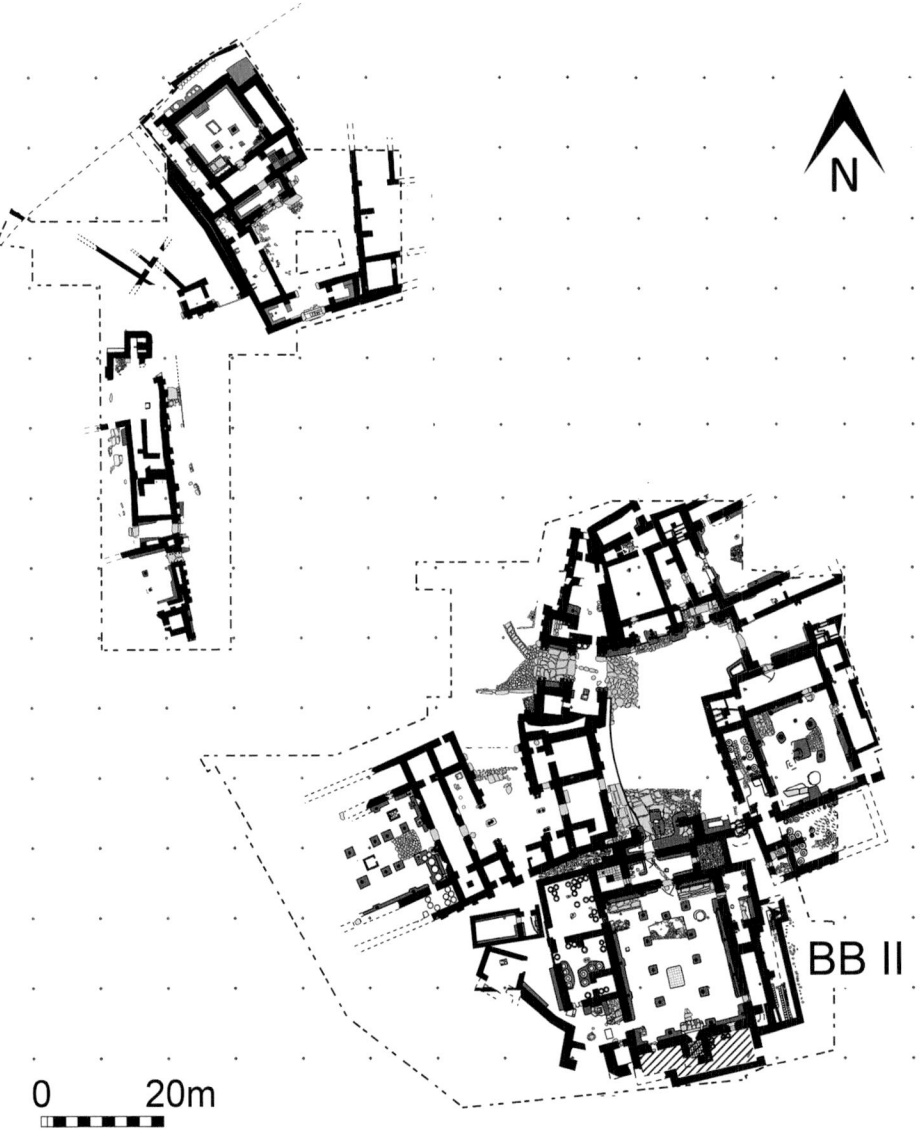

Figure 2. Site plan of the Hasanlu IVb Citadel, showing location of bead storage (Courtesy of the Penn Museum).

amplification of the status differences evident in mortuary assemblages.[2] The burials of earlier periods show few correlations between grave goods and the biological sex of the burial's occupant,[3] but in Period IVb the mortuary assemblages of men and women diverge sharply (Figure 3), illustrating a change in this community's social understanding of these male and female bodies.[4]

The burials emphasise the division of people into material categories – those who wear garments fastened by pin (biological women) and those who do not but are likely to be accompanied by weapons (biological men), indicating the increased importance of gender as a social distinction.[5] An unusual, and to my knowledge unparalleled, composite dress item was found on the chests of five women of varying ages. These burials contain the wealthiest women among the Period IVb burials, as determined by the number of metal objects, and the presence of gold, albeit in token quantities, among the dress ornaments[6] (Figure 4). The

---

[2] Cifarelli 2017.

[3] Cifarelli 2013. Osteological analyses of the partial remains of approximately 70 individuals from Period IVb burials are presented in Selinsky 2009. Selinsky's determination of the sex of the skeletons is a useful point of departure for the evaluation of sex-based patterns of artefact distribution, and helps to avoid the circular reasoning whereby the sex of a grave's occupant is inferred based on assumptions about gendered grave goods.

[4] For a brief analysis of the mortuary assemblages associated with male bodies in the Period IVb burials at Hasanlu, see Cifarelli 2016.

[5] These consistent relationships between sexed bodies and object types in Period IVb burials are evidence for what feminist philosopher Judith Butler (1993, 7-8) terms the 'regulatory schema' or rules, by which objects and people are gendered.

[6] Cifarelli in press. Four of these burials (SK59, SK448, SK484 and SK503 were excavated as part of the Hasanlu expedition and the skeletal remains have been osteologically determined to belong to women (Selinsky 2009). A fifth burial was found in Aurel Stein's 1937 excavations at Hasanlu (Stein and Andrews 1940: 397-99). Stein remarked at the time that the skeleton looked feminine, an observation borne out by the gendered grave goods.

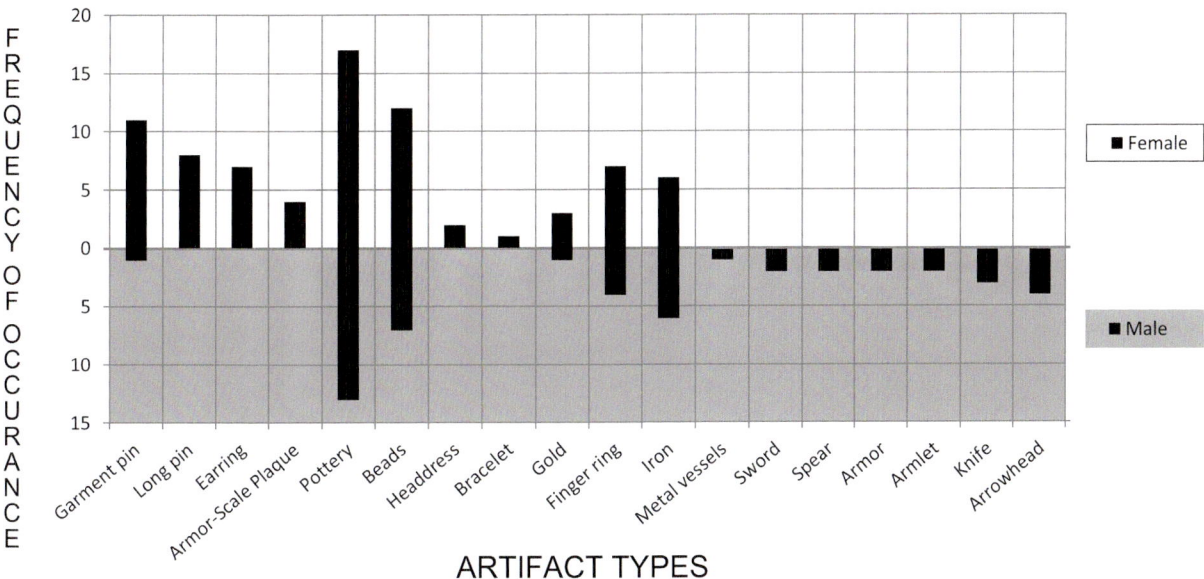

Figure 3. Graph showing frequency of occurrence of artefact types in women's and men's and women's burials during Period IVb, ranked from left to right using a z-score calculation (graph by Paul Sanchez).

Figure 4. Excavation photograph of Burial SK481, adult female, Operation VIF Burial 10 (Courtesy of the Penn Museum).

composite ornaments consist of rounded, triangular copper alloy or iron plaques that range from 7-15cm in length attached to clothing by way of riveted studs, accompanied by beads of various types, including vitreous materials, metal, carnelian and shell, as well as copper alloy tubes of varying lengths. Beaded ornaments are inherently composite in nature – they incorporate multiple elements, often in a wide range of materials, with varying methods of manufacture, geographical and chronological points of origin (Figure 5).

Beaded Dress Items

When considering beaded dress items as artefacts, it is tempting to define them in terms of individual elements, or 'parts', assembled into a 'whole'. I would argue, though, that objects composed of beads are neither parts nor whole. They are rather, in the words of Marcus Brittain and Oliver Harris, engaged in 'ongoing transformations', a cycle of creation, fragmentation and recreation.[7] The integrity of a beaded dress ornament is as fragile as the material that holds it together, and in the archaeological record these objects are very rarely found strung or sewn together in their original arrangement. Anyone who wears beaded jewellery or clothing is aware of its precarious nature, and has left at one time or another a trail of sequins or beads that if sufficiently valued are gathered up and refabricated.

These composite assemblages, therefore, require sustained engagement in the form of maintenance and

---
[7] Brittain and Harris 2010: 585.

Figure 5. Composite photograph of HAS64-193 (UM 65-31-113), based on photographs of artefacts in situ (composite by author, Courtesy of the Penn Museum).

reworking over time. As artefacts, they are analogous to Ian Hodder's example of the clay houses at Çatal Hüyük, for like clay structures, the continued existence and utility of these dress items requires frequent human interactions, by which they entangle and even entrap the humans with which they are associated.[8] Beaded jewellery was found in 40% of the burials from Hasanlu IVb, in the burials of both men and women, children, adults and the elderly. The excavation records are often imprecise, with groups of beads described as necklaces regardless of their placement. In instances where the excavators provided more explicit information, records show that beads were found at the neck, the hand or wrist, on the torso or shoulder, and in the case of certain young women, surrounding the head in a way that suggests the presence of beaded headdresses. While their precise arrangements are perhaps not possible to reconstruct, artisans seem to have combined the beads in ways that juxtapose the colours, textures, sheens, shapes and perhaps sounds of varied materials including vitreous materials, stone (often carnelian), bone, shell and metal, usually copper alloy, and very rarely antimony or gold.

A Case Study: *Arcularia* Shell Beads and Entanglement across Space and Time

As a human made object, what Ian Hodder defines as a 'thing',[9] each element in these compositions has its own point of origin and biography, and participates in broad networks of human-thing interaction across considerable expanses of time and space. Each of these objects carries its own entanglements and *chaînes opératoires*, from the extraction of raw materials; to crafting, circulation and initial use; to assembly into an item of adornment; to reassembly and reuse, and ultimately to burial, thus removal from circulation. As an example of such a thing, beginning in the Middle Bronze Age (ca. 1900–1600 BCE), a particular type of sea shell of the genus *Arcularia gibbosula* was integrated into dress ornaments found in burials at Hasanlu (Figure 6).

At Hasanlu these shells are quite a long way from their point of origin, as *Arcularia* live in the waters of the Mediterranean Sea. Their presence at Hasanlu speaks to the existence of complex networks of long distance interaction, and it isn't surprising that they appear at a time when the ceramic record at Hasanlu indicates intensive interaction with the northern Mesopotamia.[10] *Arcularia* shells are found elsewhere in the Near East, in northern Mesopotamia, Syria and the Levant,[11] as well as in the eastern Mediterranean at site in Cyprus and Greece,[12] appearing in burials, in offering contexts, and temple foundation deposits.[13] Unlike most northern Mesopotamian sites, at Hasanlu these items are not found in a single time horizon, but in contexts separated by hundreds of years. The earliest burial in which they appear at Hasanlu, Burial SK45-7, dates to the Middle Bronze Age (Hasanlu VIb, ca 1900–1600 BC), and is well furnished for the site, with high quality goods including,

---

[8] Hodder 2011: 156.
[9] Hodder 2011: 155.
[10] Danti 2013: 146-8.
[11] See McGovern and Brown 1986: map 5.
[12] Reese 1982: 86-7.
[13] Reese 1989; Reese 1992: 178.

Figure 6. Beads, including *Arcularia*, from Middle Bronze Age Burial SK45-7, Hasanlu VIb, HAS 58-134 (UM59-4-78) (Courtesy of the Penn Museum).

in addition to *Arcularia*, exceptionally well carved beads of carnelian and rock crystal.[14] *Arcularia* also appear in contexts dated to Period IVb (1050–800 BCE), including the burial of a young woman (SK481) on the low mound, and at the neck of a young adult crushed in the collapse of the largest temple at the site, BBII.[15] These shells are found as well in the treasury of Temple BBII amidst a cache of thousands of beads, perhaps integrated into beaded jewellery.[16] Once part of living creatures, then harvested from the Mediterranean sea, these shells were collected, and at some point in time were moved across long distances either as unprocessed shells, perforated for use as beads, or integrated into dress ornaments. Their distribution in the Period IVb contexts clearly demonstrates that *Arcularia* were worn as personal ornaments in life and in death, and their presence in the treasuries perhaps provide evidence for another stage in their biographies, as cultic equipment or gifts to the gods.

The Temple BBII storage context for the *Arcularia* shells was located two meters above the floor level in debris from a collapsed second story room (Figure 2, lower right). The contents, in addition to thousands of beads, included fine furniture, glassware and vases, maceheads, fine and common ceramics, metal vessels, ivory inlays, and significantly, jar sealings, which indicated that some of these valuable gifts to the deity were safeguarded.[17] Among these the finds in this context, and in other storerooms on the citadel in temples and elite residences, are numerous examples of 'heirlooms', objects whose date of manufacture considerably precedes that of the context in which they were found, a surprising occurrence given that a significant fire destroyed much of the citadel at the end of Period IVc (ca. 1050 BCE). These include Kassite glass vessels, a Middle Assyrian mace head, a bowl inscribed by a 14th or 13th century BCE Babylonian ruler, two stone maceheads inscribed with the name of the king of Susa, Tan-Ruhuarater (ca. 2100 BCE), as well as the famous Gold Bowl, likely made locally in the 11th century BCE.[18] These inscribed and otherwise datable examples provide the clearest evidence at the site for the collecting and enclaving of valuables in the citadel over time, but there can be no doubt that other older objects, perhaps more difficult to identify, will be found within the thousands of luxury objects enclaved in Hasanlu's temples and elite residences. The *Arcularia* shells in use- and storage-contexts at Hasanlu during Period IVb likely participated in this phenomenon, coming into the site during the Middle Bronze Age – the era when they first appear in burials and when the connections between Hasanlu and the west were most strong.

True heirlooms – objects transmitted from one individual to another across generations – enchain people over time, creating a liminal space where the past intersects with the present. Katina Lillios argues that the control and display of heirlooms play a significant role in constructing and reproducing elite social identity and inequality within communities.[19] We have no way of knowing if the older objects discovered in the Period IVb citadel were true heirlooms in this sense, but the elite contexts in which they were found suggest that they share the function of reinforcing local hierarchy. It seems quite clear that Hasanlu's elite collected valuable luxury goods, imported and locally made. That they did so well before Period IVb is suggested by the presence in the Period IVb destruction

---

[14] Cifarelli 2013: 314.
[15] HAS64-200 (location unknown), HAS64-421 (Teheran Museum).
[16] Reese 1989: 83.
[17] A selection of these finds were published in Muscarella 1980; de Schauensee 1988; Dyson 1989; Reese 1989; Marcus 1991; de Schauensee 2011; Danti and Cifarelli 2016. For a discussion of the sealings, see Marcus 1996: 64-5.
[18] Dyson and Pigott 1975, 183; Porada 1979; Marcus 1991.
[19] Lillios 1999: 243-4.

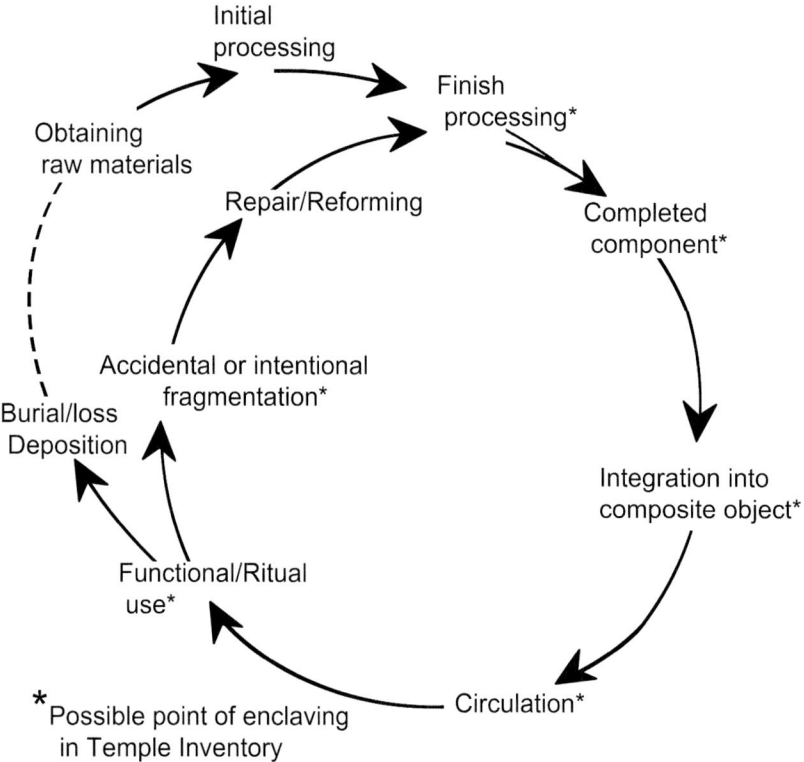

Figure 7. Object biography of beaded dress ornaments (adapted from Jennings 2014, Fig. 1).

of objects identical to those found in Period VIb burials, with proof positive provided by the presence of ivory and glazed ceramic fragments stratified below the Period IVb floor of Temple BBII.[20]

That the *Arcularia* shells are found in this range of contexts, integrated into personal ornaments in intentional as well as accidental burials and in the treasury of temple BBII, complicates their biographies as objects (Figure 7). The examples found on the citadel entered the archaeological record not through intentional deposition, but through the agency of destruction. We can't know at what stage in their object biography they were enclaved, and whether their presence there indicates that they were removed from circulation. Those found in temple treasuries may have been gifts to the gods, or cultic equipment for the use – through dress – of temple personnel. Elisa Roßberger argues that regardless of whether objects found in temples were placed there as votives or were 'temple inventory', such collections of older objects serve as the material correlate of the collective memory of the society, bringing the past into the present.[21]

These *Arcularia* shells are sufficiently rare at the site to render their wearers visually conspicuous, and their inclusion in composite dress items links their wearers both with upper reaches of the social hierarchy at the site, as well as with the past. Even if over the generations these elements had lost their specifically 'imported' or 'exotic' identity, by integrating elements which had been collected, handed down, and enclaved, into new items of adornment, the residents of Hasanlu IVb preserved the memory of, and thereby reproduced, a sense of the ancestral accomplishments and prestige that resulted in the presence of these items at the site. The example of the *Arcularia* shells, found thousands of miles from their point of origin and safeguarded for hundreds of years, illustrates the complexity of the biographies of elements that contribute to composite dress items, shows the ways these things connect the present with the past, and underscores the importance of deep history and cultural continuity over generations at the site.

Entangled by Gender: Militarisation in Period IVb

Continuity, however, is not the entire story. Period IVb – which began in the eleventh century BC with a major fire and ended with the total destruction of the site around 800 BC – was a time of change brought about by external threats and internal crises.[22] The increased emphasis on the material differentiation of men's and women's dress and mortuary assemblages is a manifestation of the social forces at work in this period. These distinctions were created using some objects that appear to have been locally made, but far more that were strongly correlated to, and perhaps imported from, the northern, often proto-Urartian material culture of sites in the South Caucasus and the Talesh.[23] These 'foreign' objects lead us back to the five burials introduced earlier, where they are integrated into composite dress items on the women's bodies. In these five burials, copper alloy and iron armour scales were decorated with beads and worn on the women's chests as dress ornaments (Figures 4, 5, 8). Armour scales are part of the kit of elite male warriors in the ancient Near East during the Late Bronze and Iron Ages.

---

[20] Cifarelli 2013; Danti and Cifarelli 2016.
[21] Roßberger 2016.

[22] Cifarelli 2017; Danti and Cifarelli 2015.
[23] Rubinson 2012; Danti and Cifarelli 2015; Cifarelli in press; 2017.

Figure 8. Excavation photograph of Burial SK 448, adult female, Operation VIC Burial 4 (Courtesy of the Penn Museum).

Archaeological and visual evidence indicates that their geographical range is extensive, from Babylonia in the south to Armenia in the north, as far west as the Mediterranean coast, with isolated examples in Greece and Cyprus.[24] The sheet metal scales are assembled into armoured garments by either lacing them together or attaching them to textile or leather,[25] and in order for them to perform their protective function they require frequent repair and reconstruction. Like jewellery elements, they are neither parts nor wholes, but objects engaged in ongoing transformations.

In addition to the five armour scales found in women's burials, approximately 37 other examples of armour scales were found at Hasanlu in Period IVb citadel contexts. They are made of iron and/or copper alloy, in a wide range of shapes, sizes and styles of decoration. Most were found in a treasury associated with Temple BBII, and two others in a storage room in elite residence BBIII. A number of examples were found in buildings and courtyards on the citadel, in what might have been use-contexts related to the battle that destroyed Hasanlu, although none were directly associated with bodies. A complete armoured garment would require a large number of similar scales, and not only are scales not found in sufficient quantities at Hasanlu, no such 'sets' occur there. Even if we assume that some of the soldiers who fell in the final battle were stripped of their armour by the victors, many combatants were struck down when buildings collapsed on them, prohibiting the looting of the dead, and none of whom were found with any scales on their bodies. There is therefore no evidence that armour scales were actually used in armoured garments by the residents of Hasanlu, although they would have been objects whose function was well known.

Of the five scales found in women's burials, only one – the iron example from Stein's excavation – has any correlates on the citadel, with one example in temple BBII and two in mixed household storage in Elite Residence BBIII. The other four examples of smaller copper alloy scales with a central rib, embossed dots around the edges, and riveted studs attaching them to textile or leather – are not found elsewhere at Hasanlu. They are paralleled, however, in slightly earlier burials at northern sites in Armenia and in the Talesh region of the Caspian coast, an area whose material culture is linked to that of the Caucasus in the Bronze and early Iron ages. Interestingly, at the site of Djonu in the Talesh, we see armour scales repurposed as women's dress, in this case linked together in a belt in what Jacques de Morgan believed to be a woman's burial.[26] The particular scales associated with women's bodies at Hasanlu, were likely northern in origin.

As was the case at many sites throughout the Near East and eastern Mediterranean, the armour scales found in the temples and residences at Hasanlu seem to have functioned synecdochally as emblems of militarism and gifts to the gods rather than useful military equipment. In his discussion of the single armour scales found in Aegean contexts, Joseph Maran describes the intellectual process by which the part stands for the whole as an 'act of abstracting' from which 'it is only one step to an apotropaic use in which a single scale is meant to convey the protective properties of the complete corselet'.[27] In the case of the armour scales in Hasanlu burials, the object undergoes an additional stage of abstraction in which they are converted and

---

[24] For a recent discussions of Near Eastern armour, see Hulit 2002, Barron 2010 and De Backer 2012. For examples from Armenia, Azerbaijan and Iran, see Morgan 1896: 47, 103; Morgan 1905: 296; Schaeffer 1948: 430-432, fig. 233.21, 30; Esayan 1990: fig. 10; Maran 2004: 18-22 (I am most grateful to Dr. Anna Paule for referring me to this article).
[25] Barron 2010: 148.
[26] Morgan 1896: 47, 103.
[27] Maran 2004: 24-5.

integrated into an entirely different sort of object. What would have been clearly recognizable to the residents of Hasanlu as a piece of elite male military equipment becomes an element of feminine attire.

These northern style elements of masculine military equipment found in the five women's burials were repurposed, converted into components of beaded dress ornaments, some of which include *Arcularia* shells. It is not possible to reconstruct the original arrangement of these components with great precision, due to the uneven quality of the excavation records. Stein describes the composite ornament as consisting of three bundles of copper alloy and iron tubes, arranged end to end and connected with copper rings, attached to and partially overlying the 'broader edge' of the armour scale. Excavation photographs of Burial SK481 suggest much the same arrangement, with two large groups of beads that appear to have been hung from the scale lying on the body's upper right chest (Figures 4, 5, 9). The beads integrated into this ornament include *Arcularia*, *Dentalium* and cowrie shells, carnelian, vitreous materials, antimony, iron, copper alloy barrel beads and various lengths of copper alloy tubes. This variety of color, sheen, size and shape provides a glimpse of the visual richness and complexity of this woman's dress. Stein further observed that when lying across 'a lady's breast [this composite object] would have vibrated as she passed and produced a pleasant musical tinkling whenever she moved'.[28] The noise-making or musical aspect of these composite artefacts is quite important, as it would have contributed significantly to their social impact.

Certainly metal scales serving their armorial function when integrated into a man's protective garment would have made distinctive sounds as the male body moved and interacted with weapons in battle, and perhaps the percussive elements were added to these dress ornaments in imitation or evocation of this effect.

There are as well numerous ethnographic and archaeological parallels for 'musical' adornment from around the world – the corded skirts decorated with similar copper tubes found in Bronze Age burials in Denmark would surely have rendered the movements of their wearers audible.[29] Both the Hebrew Bible and the Quran characterize the sounds produced by the interaction of jewellery on women's bodies as immodest,[30] indicating both its prevalence and the

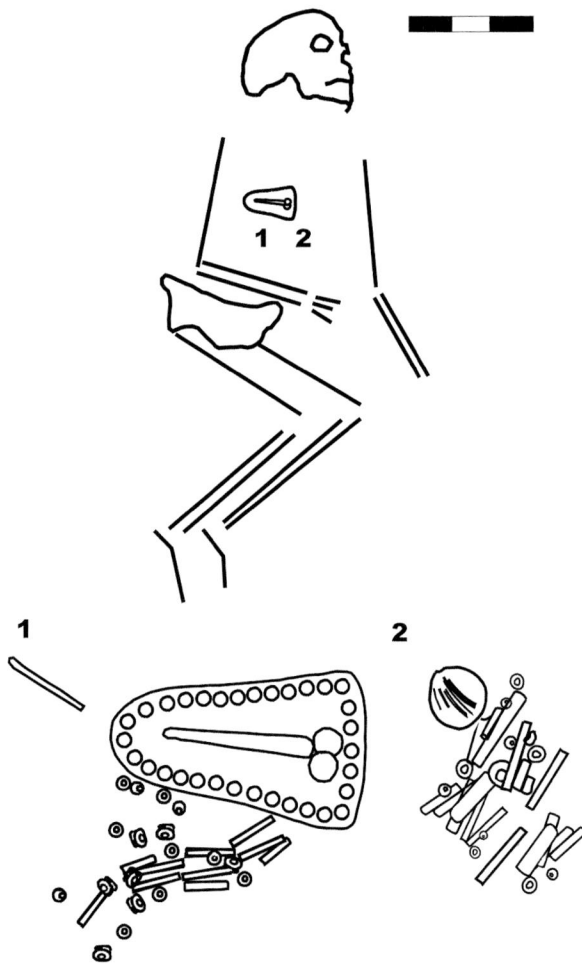

Figure 9. Excavation drawing of Burial SK481, showing location of armor scale and arrangements of bead groups that accompanied it (Courtesy of the Penn Museum).

potential nature of its appeal. The social value and constructed meaning of these composite objects, the reasons why the armour scales were converted and integrated from masculine to feminine, and from armour to adornment, present considerable challenges to interpretation. The process of fragmentation that separated the scales from their original armorial purpose may have taken place far from Hasanlu, as may the shift from their placement on men's bodies to those of women. The armour scales themselves appear to be earlier than the burials in which they are found at Hasanlu, but we cannot know if they are heirlooms in the strictest sense, linking individuals across generations and serving as a reminder of someone's social or political presence.[31] It is tempting to infer biographical information about these women and their relationships from their dress – to conclude that the integration of masculine military equipment into the dress of a few elite women manifested actual

---

[28] Stein and Andrew 1940: 397-8.
[29] Kolotourou 2007: 80-86; for examples from Denmark see Kristiansen 2013: 756, fig. 1.
[30] Surah 24: 31 of the Qur'an states 'that [women] should not strike their feet in order to draw attention to their hidden ornaments…' In the Hebrew Bible, Isa 3:16 describes the 'wanton'-eyed daughters of Zion, 'walking and mincing as they go, (make) and making a tinkling with their feet'. These examples are not intended to suggest that the

same values are in place at Hasanlu, they merely illustrate the erotic potential of audible ornamentation.
[31] Lillios 1999.

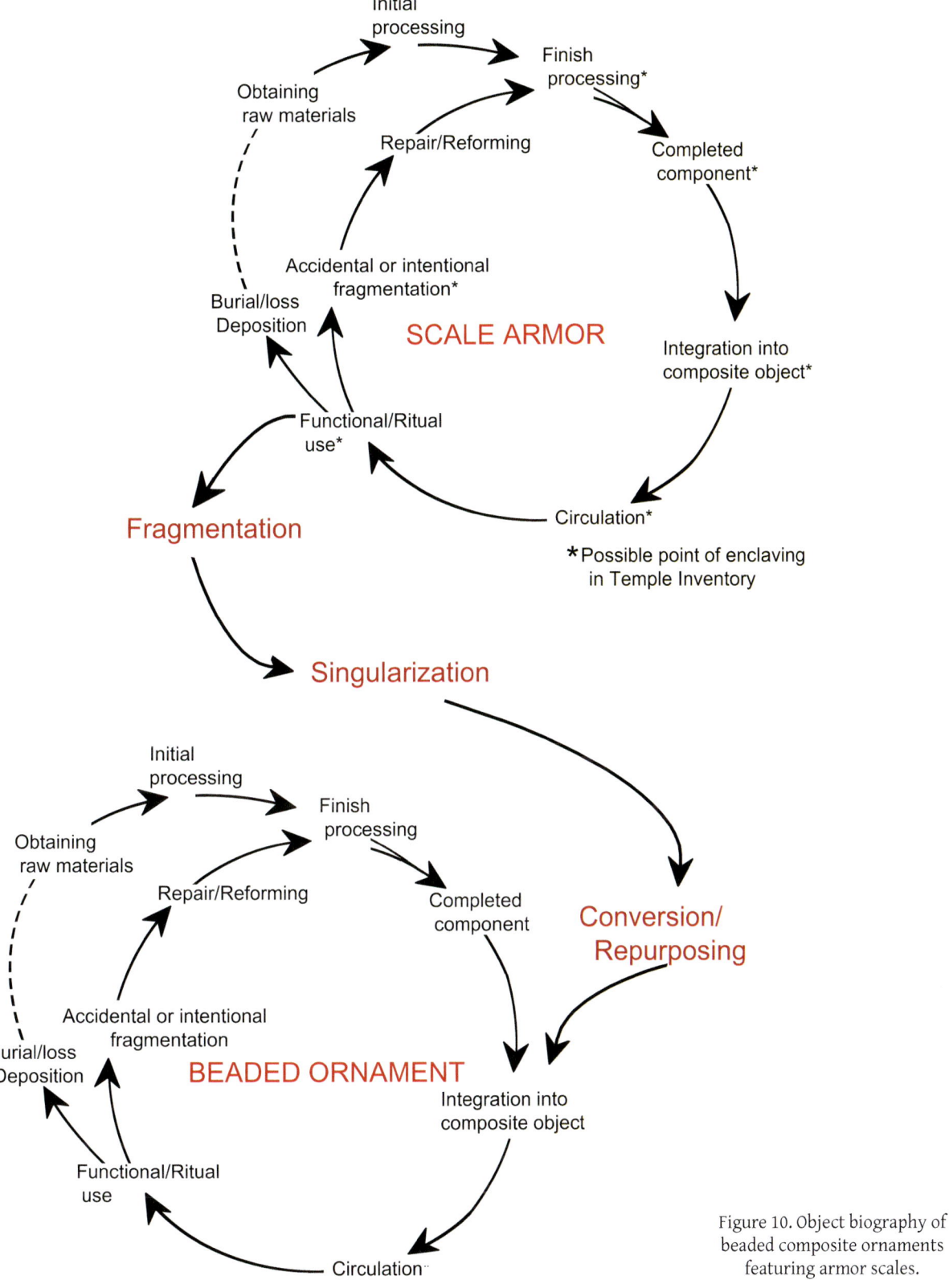

Figure 10. Object biography of beaded composite ornaments featuring armor scales.

connections or enchainments to elite military males living or dead, local or foreign. While there is ample evidence for the presence of such warriors among the Hasanlu IVb burials, the type of documentary or scientific information that could directly link individuals and support such an interpretation simply does not exist at Hasanlu.

Ian Hodder has described webs of human-thing entanglements as cables in which material, biological, social, cultural, psychological, and cognitive strands interact and bind.[32] But we simply have no way to untangle the strands that connect these artefacts on women's bodies in burials at Hasanlu to their geographical points of origin in the proto-Urartian north and original owners. Were they spoils of war? Souvenirs of travel? Gifts? Trade items? Does their presence and distribution correspond to John Chapman's notion of fragmentation and enchainment, binding individuals at Hasanlu to those in distant lands, binding these five individuals together, or binding men to women?[33] Do they function in a ceremonial or emblematic fashion on the bodies of these women in much the same role played by northern-style copper alloy armoured belts in male warrior burials?[34] Would they even have been understood at Hasanlu as 'foreign' or 'imported'? What of their original 'aura' - that which according to Jody Joy loads objects with agency, which derives in turn from their life history or biography - is retained in their resurrected existence as elements in women's composite personal ornaments?[35]

The leap of imagination by which these imported armour scales were integrated into beaded, perhaps musical, dress ornaments is remarkable. Tubes and beads were found with the armour scales in every instance, but interestingly the sets of beads associated with each of these scales are not standardized, indicating that each one of these composite objects resulted from a unique and personal human-thing interaction. The scales were separated from one arena of social value with its own biography, then inserted into another existence as part of a composite personal ornament. In its new life as part of a dress ornament, the scale retained biographical associations of its earlier existence, particularly masculinity and militarism, and perhaps foreignness. We can consider these armor scale elements as having been 'singularized', or invested with particular, personal meaning by their owners.[36] The processes of singularization allow for the conversion of this object and its integration into an entirely different, and differently gendered, type of object (Figure 10).[37] As unusual as these composite ornaments are, the series of choices by which they came into being is entirely in keeping with the cultural practices at the site that privilege the collection, curation, juxtaposition and integration of objects that are deemed special - because of their age, exotic origin, or aesthetic qualities.

This addition of a singularized armour scale - an item of great social value, esteemed highly enough to be a gift for the gods - to a woman's beaded dress ornament was an individual act, invested with rich social meaning. The visually compelling, composite ornament that linked these women across gendered and geographical space to a northern inflected, militarized masculinity, was made more prominent by the inclusion of metal tubes that would have chimed against the sheet metal scales. The attribute of audibility highlighted the movements of these women's bodies and drew attention to the bodies themselves, while contributing to the social soundscape of Hasanlu in Period IVb. These unusual, composite artefacts drew upon the rich object biographies of each constituent element, bringing the past into the present, the far into the near, the masculine into the feminine. The abundantly entangled objects, as part of the elite women's dress at Hasanlu, contributed to the construction and performance of a complex social identity, one with ties to the past, as well as links to a particularly northern, militarized masculinity.

Acknowledgements

I am very grateful to Silvana Di Paolo for including me in the ICAANE workshop on composite objects, a wonderful session that provided much food for thought and inspiration. Thanks are due as well to Michael Danti and the University of Pennsylvania Museum of Archaeology and Anthropology for permission to study and publish the Hasanlu materials. I am grateful to Joseph Maran for feedback, and as always to Paul Sanchez and Isabel Cifarelli for outstanding technical support.

References

Barron, A.E. 2010. *Late Assyrian Arms and Armour. Art versus Artifact.* (PhD Dissertation, University of Toronto, Toronto).

Brittain, M. and O. Harris 2010. Enchaining Arguments and Fragmenting Assumptions. Reconsidering the Fragmentation Debate in Archaeology, *World Archaeology* 42/4: 581-594.

Butler, J. 1993. *Bodies that Matter. On the Discursive Limits of 'Sex'.* New York: Routledge.

Cifarelli, M. 2013. The Personal Ornaments of Hasanlu VIb–Ivc. In Danti 2013: 313–323.

Cifarelli, M. 2016. Masculinities and militarization at Hasanlu. A View from the Burials. In S. Budin and J. Webb (eds), Gender Archaeology, *Near Eastern Archaeology* 79/3: 196-204.

Cifarelli, M. 2017. Archaeological Evidence for Small Scale Crisis. Hasanlu between Destructions. In J. Driessen and T. Cunningham (eds), *From Crisis to Collapse. Archaeology and the Breakdown of Social Order*: 205-231. Louvain-la-Neuve: Presses universitaires de Louvain.

Cifarelli, M. in press. Gender, Personal Adornment, and Costly Signaling in the Iron Age Burials of Hasanlu, Iran. In A. Garcia-Ventura and S. Svärd (eds), *Gender, Methodology and the Ancient Near East*. Winona Lake: Eisenbrauns.

---

[32] Hodder 2011: 164.
[33] Chapman 1996: 203–242.
[34] Danti and Cifarelli 2015.
[35] Joy 2009.
[36] As defined by Kopytoff 1986: 79–81.
[37] Jennings 2014: 174.

Chapman, J. 1996. Enchainment, Commodification, and Gender in the Balkan Copper Age, *Journal of European Archaeology* 4: 203-242.

Danti, M. 2013. *Hasanlu V. The Late Bronze and Iron I Periods*, University Museum Monograph 137 Philadelphia: University of Pennsylvania Museum Press.

Danti, M. 2014. The Hasanlu (Iran) Gold Bowl in Context. All that Glitters...., *Antiquity* 88/341: 791-804.

Danti, M. and M. Cifarelli 2015. Iron II Warrior Burials at Hasanlu, Iran, *Iranica Antiqua* 50: 61-157.

De Backer, F. 2012. Scale-Armour in the Mediterranean Area during the Early Iron Age. A) From the IXth to the IIIrd century BC., *Revue des Études Militaires Anciennes* 5: 1-38.

Dyson, R. 1989. Rediscovering Hasanlu, *Expedition* 31: 3-11.

Dyson, R. and V.C. Pigott 1975. Hasanlu, *Iran* 13: 182-85.

Esayan, S.A. 1990. Schutzwaffen aus Armenien, *Beiträge zur allgemeinen und vergleichenden Archäologie* 9-10: 83-109.

Hodder, I. 2011. Human-Thing Entanglement. Towards an Integrated Archaeological Perspective, *Journal of the Royal Anthropological Institute* 17/1: 154-177.

Hulit, T.D. 2002. *Late Bronze Age Scale Armour in the Near East: an Experimental Investigation of Materials. Construction and Effectiveness, with a Consideration of Socio-Economic Implications* (PhD Dissertation, Durham University, Durham).

Jennings, B. 2014. Repair, Recycle or Re-use? Creating Mnemonic Devices through the Modification of Object Biographies during the Late Bronze Age in Switzerland, *Cambridge Archaeological Journal* 24/1: 163-176.

Joy, J. 2009. Reinvigorating Object Biography. Reproducing the Drama of Object Lives, *World Archaeology* 41/4: 540-556.

Kolotourou, K. 2007. Rattling Jewellery and the Cypriot Coroplast, *Archaeologia Cypria* V: 79-99.

Kopytoff, I. 1986. The Cultural Biography of Things. Commoditization as a Process. In A. Appadurai (ed.), *The Social Life of Things. Commodities in Cultural Perspective*: 64-91. Cambridge: Cambridge University Press.

Kristiansen, K. 2013. Female Clothing and Jewelry in the Nordic Bronze Age. In S. Bergerbrant and S. Sabatini (eds), *Counterpoint. Essays in Archaeology and Heritage Studies in Honour of Professor Kristian Kristiansen*, BAR International Series 2508: 755-769. Oxford: Archaeopress.

Lillios, K.T. 1999. Objects of Memory. The Ethnography and Archaeology of Heirlooms, *Journal of Archaeological Method and Theory* 6/3: 235-262.

Maran, J. 2004. The Spreading of Objects and Ideas in the Late Bronze Age Eastern Mediterranean. Two Case Examples from the Argolid of the 13th and 12th centuries BCE, *Bulletin of the American School of Oriental Research* 336: 12-30.

Marcus, M.I. 1991. The Mosaic Glass Vessels from Hasanlu, Iran. A Study in Large-Scale Stylistic Trait Distribution, *The Art Bulletin* 73/4: 536-560.

Marcus, M. 1996. *Emblems of Identity and Prestige. The Seals and Sealings from Hasanlu, Iran.* Hasanlu Special Studies 3. Philadelphia: The University Museum.

McGovern, P. and R. Brown 1986. *Late Bronze and Early Iron Ages of the Central Transjordan. The Baq'ah Valley Project, 1977-1981*, University Museum Monograph 65. Philadelphia: The University Museum.

de Morgan, J. 1896. *Mission scientifique en Perse. Tome quatrième, recherches archéologiques. Première partie.* Paris: E. Leroux.

de Morgan, J. 1905. *Recherches au Talyche 1901.* Mémoires de la Délégation en Perse 8. Paris: E. Leroux.

Muscarella, O.W. 1980. *The Catalogue of Ivories from Hasanlu, Iran*, University Museum Monograph 40 Philadelphia: The University Museum.

Reese, D. 1982. Marine and Freshwater Molluscs from the Epipaleolithic site of Hayonim Terrace, Western Galilee, Northern Israel, and Other East Mediterranean Sites, *Paléorient* 8/2: 83-90.

Reese, D. 1989. Treasures from the Sea. Shells and Shell Ornaments from Hasanlu, *Expedition* 31: 80-6.

Reese, D. 1992. Shells from the Hoard at Khirbet Karhasan (IRAQ). Appendix A. In D. Tucker, *A Middle Assyrian Hoard from Khirbet Karhasan, Iraq, Iraq* 54: 178-81.

Roßberger, E. 2016. E. Dedicated Objects and Memory Construction at the Ištar-Kitītum Temple at Iščāli. In R. Stucky, O. Kaelin and H. Mathys (eds), *Proceedings of the 9th International Congress on the Archaeology of the Ancient Near East, Basel, June 9-13, 2014, vol. 1*: 419-430. Wiesbaden: Harrasowitz.

Rubinson, K. 2012. Urartian (?) Belts and Some Antecedents. In S. Kroll, C. Gruber, U. Hellwag, M. Roaf and P. Zimansky (eds), *Biainili-Urartu. The Proceedings of the Symposium held in Munich 12-14 October 2007*: 391-96. Leuven: Peeters.

de Schauensee, M. 1988. Northwest Iran as a Bronzeworking Center. The View from Hasanlu. In J. Curtis (ed.) *Bronzeworking Centres of Western Asia c. 1000-539 B.C.*: 45-56. New York and London: Kegan Paul International and the British Museum.

de Schauensee, M. (ed.) 2011. *Peoples and Crafts in Period IVB at Hasanlu Tepe, Iran.* University Museum Monograph 132. Philadelphia: The University Museum.

Selinsky, P. 2009. *Death a Necessary End. Perspectives on Paleodemography and Aging from Hasanlu*, Iran (PhD Dissertation, University of Pennsylvania, Philadelphia).

Shaeffer, C.F.A. 1948. *Stratigraphie comparée et chronologie de l'Asie Occidentale.* London: Oxford University Press.

Stein, A. 1940. *Old Routes of Western Iran. Narrative of an Archaeological Journey Carried Out and Recorded.* New York: Greenwood Press.

Section 3

Sum of Fragments, Sum of Worlds

Composing Figural Traditions in the Mesopotamian Temple

Jean M. Evans
Oriental Institute Museum, University of Chicago
jmevans@uchicago.edu

Abstract

Two fragmentary statues of bull men dating to the beginning of the Early Dynastic period likely would have been completed with prestige stones and metals and follow a composite tradition already established in earlier Uruk theriomorphic sculpture. The use of prestige stones and metals for composite sculpture is also attested in the human figural tradition of Early Dynastic temple sculpture by a small number of surviving examples. Regardless of whether an Early Dynastic temple statue was a composite assemblage of prestige materials or whether it was made of white stone(s) of comparatively local availability, however, the frequency of drilled holes nevertheless reveals that the human figure was assembled in parts – usually head, body, and feet. The majority of Early Dynastic temple sculpture therefore belongs to the composite sculpture tradition. In addition, body parts – primarily the head and foot – were manufactured separately, usually as pendants, throughout the 3rd millennium BCE. Often associated with temples, this category of artefact is an important but overlooked aspect of the Mesopotamian figural tradition. The manufacture of body parts for both temple sculpture and as pendants in the 3rd millennium BCE has a conceptual relationship with the life-sized clay body parts and then small clay figurines of individuals pointing to or grasping body parts in the second millennium BCE. Despite shifts in material, scale, and potentially also function, these connections attest to the lengthy traditions underlying the material methods for presencing the human form.

Keywords: Early Dynastic, sculpture, temple, composite, figural, body, prestige

Figure 1. Early Dynastic bull man
(George Ortiz collection; photo by author).

The earliest surviving inscribed statue is a standing, belted bull men recording what is likely an Early Dynastic dedication (Figure 1).[1] Thought to be one of a pair, the bull men follow a composite tradition already established in earlier Uruk theriomorphic sculpture.[2] Both bull men are carved from a yellow-hued stone with translucent qualities.[3] The eyes, eyebrows, hair, beard, horns, tail, lower legs, and perhaps also the ears of the bull men would have been constructed of different, likely prestige, materials and attached separately.[4] A vertically-drilled hole in the top of the head of each bull man suggests they were meant as supports for another item.[5]

The use of prestige stones and metals is also attested in the human figural tradition of Early Dynastic temple sculpture by a small number of surviving examples. The composite bull men therefore are not merely the vestiges of an earlier tradition. That is, the bull men provide a link from the theriomorphic sculpture that dominated the earlier Uruk composite sculpture corpus

---

[1] The bull men are said to be from Ummar and were first published by Frankfort 1939: no. 206 and Lloyd 1946.
[2] Lloyd (1946: 3-4) notes differences between the two figures but concludes it is 'not impossible' that they are indeed a pair.
[3] Both Le Breton (1957: 109) and Moorey (1994: 24) suggest a source in the Zagros Mountains for such translucent stones.
[4] Lloyd (1946: 3-4) notes differences in the arrangement of the drilled holes among the two bull men but suggests similar reconstructions.
[5] See, for example, Frankfort 1939: nos. 181-183.

to later composite sculpture among Early Dynastic temple sculpture of human figures with clasped hands. It is therefore noteworthy that the bull men are literally part animal and part human but clasp their hands in the manner of a temple dedicant.

In the Early Dynastic sculpture hoard from the Abu Temple at Tell Asmar, the kneeling belted figure, which also functions as a support, literally wearing a vessel as a hollowed-out headdress, is also carved from a yellow-hued stone with translucent qualities.[6] Some Early Dynastic sculpture fragments from other Diyala temples are – like the pair of bull men and the Asmar hoard kneeling belted figure – also carved from a yellow-hued stone with translucent qualities.[7] One example is a torso fragment from the Shara Temple preserving the beginning of a smooth skirt with a band at the waist and a tassel at the back, on the left (Figure 2). The remains of a skirt reveal that the fragment represents a human figure. Because there is no break in the stone above the waistband, however, it is unlikely that the hands had been clasped – as is also suggested by the angle of the preserved upper left arm. The figure therefore may have served an additional support function, perhaps assuming an action pose like the belted male figure supporting a vessel also from the Shara Temple.[8] A crouching nude male figure bearing a load on the back, either Sin Temple VI or VII at Khafajah, attests to such poses among human subjects.[9] Although representing a human figure – rather than a mythological figure, such as a bull man or belted male figure – the Shara Temple sculpture fragment (Figure 2) is a composite statue. The beard is carved in relief as a rectangle against the chest with a band in lower relief along the sides and bottom. The rectangle of the beard is smooth, but the surrounding band in lower relief has a rough surface, suggesting it had been fitted with a different material. A drilled hole in the right arm may also be indicative of another attachment – or simply a repair. As with the pair of bull men, prestige stones and metals had probably completed the composition.

Figure 2. Tell Agrab, Shara Temple, Early Dynastic sculpture fragment of a male figure (Ag. 35:999, by permission of the Oriental Institute).

The surviving inscription on one of the pair of bull men is difficult to read but appears to record a royal dedication by a ruler of Umma.[10] The bull men and belted heroes under discussion here therefore might be understood as constituting a category of royal dedication at the beginning of the Early Dynastic period.[11] It is therefore interesting to note that the other bull man of the pair had originally been inscribed, and the inscription 'had been broken off with a precision which could be deliberate'.[12] Similarly, the eyes of the Shara Temple belted male figure supporting a vessel are gouged out.[13] These mutilations parallel the fate of royal representation known from both textual sources and material remains.[14] Later in the Early Dynastic period, rulers dedicated statues representing themselves rather than mythological bull men. Instead of light-coloured stones with translucent qualities, dark stones were subsequently preferred for royal representation in southern Mesopotamia. Once temple sculpture of human figures with clasped hands comes to dominate Early Dynastic sacred gifting practices, white stones – most often identified as gypsum although seldom subjected to scientific analysis – dominate the corpus. Composite sculpture of prestige stones and metals

---

[6] Frankfort 1939: 25, no. 16.
[7] Frankfort 1943: nos. 262, 272, 293, Ag. 36:491.
[8] Frankfort 1943: no. 269.
[9] Frankfort 1939: no. 92.

[10] See, most recently, Marchesi 2004: 196-197.
[11] One of the copper supports from the Temple Oval has an illegible inscription – probably dedicatory – on the back (Frankfort 1939: 12, no. 181).
[12] Lloyd 1946: 3; see also Woods 2012: 38-39.
[13] Frankfort 1943: pl. 34B.
[14] For the Early Dynastic period, see Woods 2012.

nevertheless is well attested in textual sources.[15] A few rare examples of composite temple sculpture have also survived.[16]

Like the supports discussed above, a standing female figure from the Inanna Temple is a composite statue carved from a coloured stone with translucent qualities (Figure 3).[17] Instead of a yellow-hued stone, the stone is green. Translucent green stones, variously described as serpentine, gypsum, or calcite – none has been subject to scientific analysis – are attested among earlier stamp seals and theriomorphic pendants as well as among contemporary glyptic.[18] Although the Inanna Temple statue is the only surviving example of the use of this green stone for temple statues of human donors, green-hued stones with translucent qualities are attested among vessels, theriomorphic sculpture, and mace heads from, for example, the Shara Temple at Tell Agrab.[19] The Inanna Temple statue had been buried, along with other sacred gifts, in a bench in cella 179 of level VIIB. It is among the smallest of the statues excavated from the Inanna Temple, perhaps reflecting the expense of its materials. The portion of the statue carved from green stone is the body. The head is missing except for the gold face with inlaid eyes of shell and lapis lazuli affixed with bitumen. Shortly after excavation, traces of wood were observed adhering to the gold face, suggesting that the head was originally of wood, plated with gold, and attached to the body by the hole drilled into the green stone at the neck. The feet and base are also missing, but two drilled holes appearing to preserve traces of bitumen are present on the underside of the green stone. The excavators associated two pieces of wood found near the bottom of the statue with the remains of its base. Perhaps the base had also been plated, like the face, with gold or another prestige metal.

While the Inanna Temple composite statue signifies elite artistic production with its use of prestige materials, this composite statue – with its drilled holes at the neck and underside to attach the head and feet separately – nevertheless recalls in manufacturing technique the common method for assembling Early Dynastic temple statues. Among the surviving examples, prestige composite materials, most often lapis lazuli, were usually limited to eyes and eyebrows and, more rarely, the nipples and parts of the hair. However, regardless of whether an Early Dynastic temple statue

Figure 3. Nippur, Inanna Temple VIIB, Early Dynastic statue of a standing female figure (7N-184, by permission of the Nippur Publication Project).

was a composite assemblage of prestige imported materials or whether it was made of white stone(s) of comparatively local availability, the frequency of drilled holes located most commonly at the neck for the attachment of the head and the underside of the skirt for the attachment of the feet/base reveals that the human figure was usually assembled from parts – head, body, and feet/base (Figures 4-5). In the Early Dynastic temple sculpture tradition, the frequency of drilled holes on the core stone body indicates that the addition of head and feet/base was a component of the original construction of the sculpture and therefore a fundamental means of assembly. The majority of Early Dynastic temple sculpture is therefore composite.

The manufacture of heads and feet separately, most often as pendants, might suggest there is a significance beyond facilitating manufacture in conceiving of heads and feet as independent parts. Head pendants and foot pendants are the dominant body parts represented among Mesopotamian pendants, and few other body parts are ever represented independently. These pendants form an important category of artefact and constitute an overlooked aspect of the Mesopotamian figural tradition. For example, a series of head pendants, usually made of shell, are pierced laterally, ear-to-ear, for suspension (Figure 6). These pendants have stylistic parallels with Early Dynastic temple sculpture and similar inlaid eyes attached by bitumen,

---

[15] Archi 1999; 2005; Marchesi 2004.
[16] See also Matthiae 2009 for the two composite female figures from the Administrative Quarter in Palace G at Ebla that appear to have been part of a larger ceremonial object.
[17] The statue will be published fully in Zettler and Wilson forthcoming; see Hansen 1975: no. II.
[18] Moorey 1994: 24; see also Frankfort 1935: 21-23, fig. 22; Delougaz and Lloyd 1942: Ag. 35:658, Ag. 35:121.
[19] Delougaz and Lloyd 1942: Ag. 35:1104, Ag. 35:1124; Frankfort 1943: no. 304.

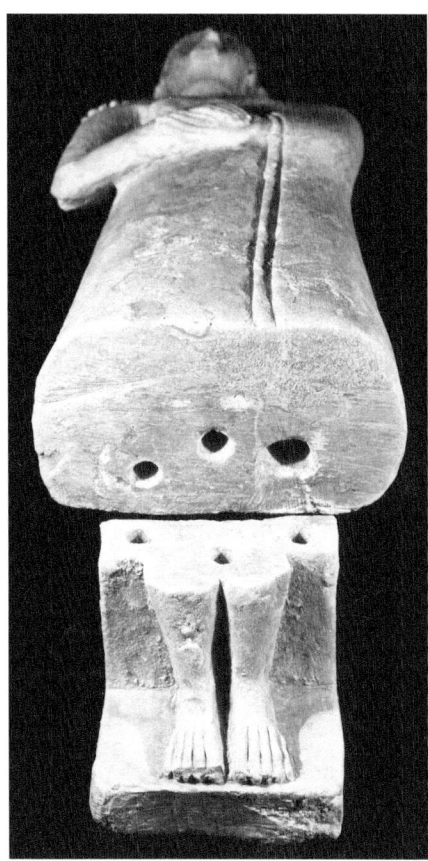

Figure 4. Nippur, Inanna Temple VIIB, Early Dynastic statue of a standing female figure (7N-162, by permission of the Nippur Publication Project).

Figure 5. Tell Asmar, Abu Temple, Early Dynastic statue of a standing male figure (As. 33:439, by permission of the Oriental Institute).

Figure 6. Khafajah, Small Temple, Early Dynastic pendant of a human head (Kh V 24, by permission of the Oriental Institute).

which is also used to articulate the eyebrows and hair. Many such pendants are concentrated in temples in the Diyala region, but pendants of the human head have a long history and are attested throughout Mesopotamia with variations.[20] In general, there is a disproportionate number of abandoned heads in Early Dynastic temples.[21] The removal of the head from the body evokes long-established pratices in which the heads of clay figurines are systematically broken from the body. In prehistoric contexts yielding also administrative remains, the breaking of clay figurines has been assigned an associated function, while later texts correlate the destruction of a clay figurine with the destruction of its potency. It should be recalled that the temple statue itself has been interpreted as a contractual obligation with the temple, with the continued benevolence of the deity in proportion to the continued maintenance of the statue.[22] The disassembly of the head of the temple statue might be understood as ending this obligation. Perhaps head pendants are somehow also related to establishing, witnessing, or completing temple transactions that ultimately sought the benevolence of the deity.

The manufacture of separate body parts allows us to connect the Mesopotamian figural tradition of the 3rd millennium BCE to 2nd millennium BCE life-sized clay body parts and then small clay figurines of individuals pointing to or grasping body parts. Despite shifts in material, scale, and potentially also function, this connection attests to the lengthy traditions underlying the material methods for presencing the human form. Baked clay sculpture of primarily life-sized female figures was reportedly looted from the site of Isin and subsequently acquired by European and American museums in the 1970s.[23] Attested are baked-clay female heads, which are often pedestaled and therefore were not fragments of larger statues (Figure 7). It is possible

[20] For example, see Delougaz and Lloyd 1942: Ag 35:813, Kh IV 153, Kh IV 164, Kh IV 187, Kh IV 209, Kh V 24.
[21] Evans 2012: 137–143.
[22] Gelb 1987: 138.
[23] Abada 1974; Blocher 1987: 30–36; Moorey 2005: 117–118, no. 157.

Figure 7. Old Babylonian baked clay sculpture of a female head (by permission of the Boston Museum of Fine Arts).

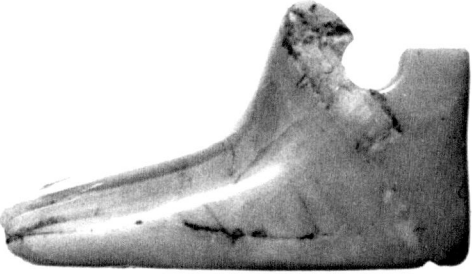

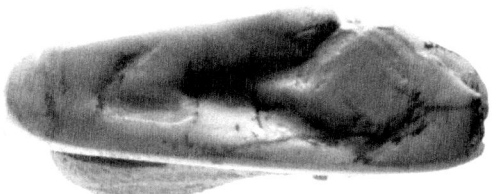

Figure 8. Tell Agrab, Shara Temple, Early Dynastic pendant of a human foot with scorpion incised on sole of the foot (Ag. 35:708, by permission of the Oriental Institute).

Figure 9. Al-Hiba, Area B, baked clay foot (after Hansen 1970, fig. 18).

that the practice of manufacturing heads separately – both for attachment to the body in Early Dynastic temple sculpture as well as independently as a category of pendant – is somehow connected to the manufacture of these later pedestalled baked clay heads. But the shift is radical. Not only is the scale life-sized in the 2nd millennium BCE, but the presence of the pedestal is jarring in its implied acceptance of the fragment as the final product. In the Early Dynastic temple sculpture tradition, it is always assumed that head and feet are destined for attachment – to make the fragment whole – as is suggested by the presence of drilled holes.

Like heads, feet also have a long tradition of being represented independently of the body. Some prehistoric stamp seals are carved with a composition of feet and either a serpent or scorpion, which would appear to connote an apotropaic quality. Early Dynastic foot pendants sometimes have scorpion imagery carved into the bottom, again evoking an apotropaic quality (Figure 8). In another chronological leap into the second millennium BCE, nearly life-sized, baked clay feet were represented independently (Figure 9). The preponderance of feet among the assemblage of nearly life-sized baked clay sculpture at Isin suggests these body parts were not always the fragments of larger statues.[24] Spycket studied this corpus of material firsthand in the Iraq Museum and observed that some of the unpublished examples appeared to have been finished off above the ankle, rather than broken off.[25]

Similarly, baked clay feet with smoothed off ends and one example which appears to have been fashioned entirely in the round were observed among the assemblage of nearly life-sized baked clay sculpture at Mashkan-Shapir, retrieved from the surface of the mound.[26]

The best-stratified context for life-sized baked clay sculpture is a foot preserving only the toes retrieved from an Old Babylonian level of the Gula Temple in area WA at Nippur.[27] A later level yielded figurines who appear to draw attention to afflicted body parts and are similar to those found near 'Aqar Quf that are associated with dog figurines and dedications to Gula (Figure 10).[28] The

---

[24] Hrouda 1977: 40-41, pl. 8; 1992: IB1814.
[25] Spycket 1990: 80.
[26] Stone and Zimansky 2004: 97-8, figs. 57-58, although the authors interpreted the smoothed-off ends 'as though the limb was originally made of several pieces and fitted together over an internal framework, perhaps of wood'.
[27] Gibson 1978: 11-12, fig. 17.2, interpreted as belonging to a life-sized statue.
[28] Mustafa 1947.

Figure 10. Nippur, Area WA, Kassite clay figurine (19N-142, by permission of the Nippur Publication Project).

excavated examples of baked clay sculpture from Isin are also associated with Gula and perhaps specifically with one of the ramps to her temple.[29] Whether all life-sized baked clay body parts should be associated with Gula is an open question, although a general connection to temples is consistent. The Mashkan-Shapir assemblage of baked clay sculpture appears to have been localised in the area of a temple platform.[30] At al-Hiba, one nearly life-sized baked clay foot was found in the top fill of Area B, the Bagara of Ningirsu.[31] The sequence in the Gula Temple at Nippur suggests that baked clay body parts in the early 2nd millennium BCE might be related to – and might have been supplanted by – figurines thought to indicate, through gesture, ailments. It therefore must be asked whether the life-sized baked clay heads and feet fulfilled a function similar to the later figurines. The Nippur figurines as well as comparable figurines found near 'Aqar Quf have been described as grasping, in addition to the stomach or back, the throat, chin, eyes, and so forth. These gestures, as well as associations with Gula, have led to the suggestion that the figurines represent individuals with afflictions from which they wish to be cured.[32] A particular emphasis on gestures which draw attention to the head, however, might signify an all-encompassing healing or well-being rather than afflictions associated with a specific body part. The question, then, is whether the head was representative of the whole body, as it also may have been when pedestalled as an independent body part, when suspended as a pendant, and when separated from the body of a temple statue in order to

end some aspect of obligation represented by the act of dedication.

In summary, much Early Dynastic temple sculpture is composite regardless of whether the raw materials of composition were prestige stones and metals or white stone(s) of relative local availability. The manufacture of independent head and foot pendants, which have a lengthy history, parallel the manufacture of independent heads and feet for Early Dynastic temple statues.

The body parts which appear to figure prominently among the corpus of human figural imagery during the early 2nd millennium BCE cannot be understood without the lengthy tradition which preceded them, and the subsequent human figurines gesturing to body parts also belong to this tradition. The conservative quality of Mesopotamian temple traditions therefore allows us to think about the declining popularity of stone temple statues after the Early Dynastic period as at least in part reflecting new ways of figuring the body as itself a composite construction and thus the representation of individual elements of the composite tradition – head and feet – became prominent. Ultimately, Early Dynastic temple sculpture belongs to a larger corpus of human figural imagery, and the composite elements of such sculpture as well as the related pendants discussed above are the ancestors of the later phenomenon of representing body parts independently – among baked clay sculpture of the Old Babylonian period – and highlighting body parts through gesture – among the clay figurines of the Kassite period.

References

Abada, K.M. 1974. Objects Acquired by the Iraq Museum 4 (Arabic), *Sumer* 39: 329-334.

Archi, A. 1999. The Steward and his Jar, *Iraq* 61: 147-158.

Archi, A. 2005. The head of Kura – the head of 'Adabal, *Journal of Near Eastern Studies* 64: 81-100.

Blocher, F. 1987. *Untersuchungen zum Motiv der nackten Frau in der altbabylonischen Zeit*, Münchener Vorderasiatische Studien Band 4. Munich: Profil Verlag.

Böck, B. 2014. *The Healing Goddess Gula: Towards an Understanding of Ancient Babylonian Medicine*, Culture and History of the Ancient Near East 67. Leiden: Brill.

Delougaz, P. and S. Lloyd 1942. *Presargonid Temples in the Diyala Region*, Oriental Institute Publications 58. Chicago: The Oriental Institute.

Evans, J.M. 2012. *The Lives of Sumerian Sculpture. An Archaeology of the Early Dynastic Temple*. Cambridge: Cambridge University Press.

Frankfort, H. 1935. *Oriental Institute Discoveries in Iraq, 1933/34. Fourth Preliminary Report of the Iraq Expedition,*

---

[29] Hrouda 1977: 31.
[30] Stone and Zimansky 2004: 349-350.
[31] Hansen 1970: fig. 18.
[32] Gibson 1990: 22; Mustafa 1947: 20; Böck 2014: 114-115.

Oriental Institute Communications 19. Chicago: The Oriental Institute.

Frankfort, H. 1939. *Sculpture of the Third Millennium B.C. from Tell Asmar and Khafājah*, Oriental Institute Publications 44. Chicago: The Oriental Institute.

Frankfort, H. 1943. *More Sculpture from the Diyala Region*, Oriental Institute Publications 60. Chicago: The Oriental Institute.

Gelb, I.J. 1987. Compound Divine Names in the Ur III period. In F. Rochberg-Halton (ed.), *Language, Literature, and History. Philological and Historical Studies Presented to Erica Reiner,* American Oriental Series 67: 125-138. New Haven: American Oriental Society.

Gibson, M. 1978. *Excavations at Nippur. Twelfth Season*, Oriental Institute Communications 23 Chicago: The Oriental Institute.

Gibson, M. 1990. Nippur, 1990. The Temple of Gula and a Glimpse of Things to Come. In W. Sumner (ed.), *The Oriental Institute 1989-1990 Annual Report*: 17-26. Chicago: The Oriental Institute.

Hansen, D.P. 1970. Al-Hiba, 1968-1969, A Preliminary Report, *Artibus Asiae* 4: 143-158.

Hansen, D.P. 1975. Frühsumerische und frühdynastische Rundplastik. In W. Orthmann (ed.), *Der Alte Orient*, Propyläen Kunstgeschichte 14: 158–170. Berlin: Propyläen Verlag.

Hrouda, B. (ed.) 1977. *Isin - Išān Bahrīyāt I. Die Ergebnisse der Ausgrabungen 1973-1974*, Bayerische Akademie der Wissenschaften. Philosophisch-Historische Klasse Abhandlungen. Neue Folge, Heft 79, Veröffentlichungen der Kommission zur Erschliessung von Keilschrifttexten. Serie C/2. Stück. Munich: Verlag der Bayerischen Akademie der Wissenschaften.

Hrouda, B. (ed.) 1992. *Isin - Išān Bahrīyāt IV. Die Ergebnisse der Ausgrabungen 1986-1989*, Bayerische Akademie der Wissenschaften. Philosophisch-Historische Klasse Abhandlungen. Neue Folge, Heft 105, Veröffentlichungen der Kommission zur Erschliessung von Keilschrifttexten. Serie C/5. Stück. Munich: Verlag der Bayerischen Akademie der Wissenschaften.

Le Breton, L. 1957. The Early Periods at Susa, Mesopotamian Relations, *Iraq* 9: 79-124.

Lloyd, S. 1946. Some Recent Additions to the Iraq Museum, *Sumer* 2: 1-9.

Marchesi, G. 2004. Who was Buried in the Royal Tombs of Ur? The Epigraphic and Textual Data, *Orientalia* 73: 153-97.

Matthiae, P. 2009. The Standard of the *maliktum* of Ebla in the Royal Archives Period, *Zeitschrift für Assyriologie* 99: 270-311.

Moorey, P.R.S. 1994. *Ancient Mesopotamian Materials and Industries*. Oxford: Clarendon Press.

Moorey, P.R.S. 2005. *Ancient Near Eastern Terracottas. With a Catalogue of the Collection in the Ashmolean Museum, Oxford*. Oxford: Ashmolean Museum.

Mustafa, M.A. 1947. Kassite Figurines. A New Group Discovered near 'Aqar Quf, *Sumer* 3: 19-22.

Spycket, A. 1990. Ex-voto mésopotamiens du IIe millénaire av. J.-C. In O. Tunça (ed.), *De la Babylonie à la Syrie, en passant par Mari. Mélanges offerts à Monsieur J.-R. Kupper à l'occasion de son 70e anniversaire*: 79-86. Liège: Université de Liège.

Stone, E. and P. Zimansky 2004. *The Anatomy of a Mesopotamian City: Survey and Soundings at Mashkan-shapir*. Winona Lake: Eisenbrauns.

Woods, C. 2012. Mutilation of Image and Text in Early Sumerian Sources. In N. May (ed.) *Iconoclasm and Text Destruction in the Ancient Near East and Beyond*, Oriental Institute Seminars 8: 33-56. Chicago: The Oriental Institute.

Zettler, R.L. and K.L. Wilson forthcoming. *Nippur VI. The Inanna Temple*, Oriental Institute Publications. Chicago: The Oriental Institute.

Polymaterism in Early Syrian Ebla

Frances Pinnock
'Sapienza' University of Rome
frances.pinnock@uniroma1.it

Abstract

While in Early Dynastic Mesopotamia polymaterism was not such a diffused practice, at Ebla, the use to compose figures with different materials, in large, as well as in miniature size images reached a peak, in the skills manifested in the assemblage of the materials, in the wealth of the raw materials employed and in the variety of objects produced. The evidence provided by this important Early Syrian town will be analysed in detail, and an attempt will be made, also with the aid of the cuneiform evidence, at understanding the reasons for the use of different materials, and the eventual symbolism behind them. Though the palace furniture was found in scattered pieces, as a consequence of the sack following to the fall of the town, the reconstruction of some unique object is proposed, which was peculiar to the Early Syrian culture of Ebla.

Keywords: Ebla, Royal Palace G, Early Syrian period, image of power, Palace furniture

Our perception of ancient art is frequently biased because it is based on the state of preservation of the artefacts as they reached us: we have a very limited amount of objects, as compared to the amount really produced, and we mostly have incomplete artefacts. Sometimes we are not even able to understand which was their original aspect. It is well known that, in our appreciation for Greek and Roman statues, the perception of the quality of the marble or bronze from which they were composed still plays a major role. The aspect of these statues, as is reconstructed today with their original colours, may look quite distant from the general ideal of classical harmony as was elaborated in the past, which led to a merely aesthetic appreciation of their beauty.[1]

A similar phenomenon may be observed in ancient Near Eastern art, where painting was used in order to cover the surfaces, or to mark some detail, in statues and relief.[2] The use of different materials in the composition of an artefact may in some measure provide indications about the final aspect of some objects. Yet, this does not solve the problem of the perception of works of art by their contemporaries: which was the real reason for the use of different colours, which aesthetic, or general ideological, principle was behind the choice for specific materials and/or specific colours, besides the certainly possible, and in part at least determinant, wish to make the representations of living beings and vegetable elements as much realistic as possible, though we should also perhaps understand what 'realistic' might mean in those regions and times.[3]

In Early Dynastic Mesopotamia polymaterism is not such a diffused practice, and is frequently limited to the application of eyes in human statues, or of stone elements representing colour patches in animal fur,[4] whereas it is more customary for the production of pieces of furniture, as proved most of all by the findings of the Royal Cemetery of Ur, and of inlaid wall panels. At Ebla, on the contrary, the use to compose figures with different materials, in large, as well as in miniature size images reaches a peak, in the skills manifested in the assemblage of the materials, in the wealth of the raw materials employed and in the variety of objects produced.[5] Many scattered fragments of furniture and

---

[1] Particularly in Italy, archaeological pieces, specially statues, were frequently used for the adornment of private houses since immediately after the fall of the Roman Empire. Members of aristocratic families, when building their villas and palaces, frequently found fragments of, or complete Roman statues, which were placed inside their houses, or even in the gardens, as beautiful pieces of furniture, disregarding their value, and obviously oblivious of their relation with their original contexts.

[2] See, recently, the contributions in the Session 'Colour and Light in Architecture, Art and Material Culture' in the 7th ICAANE of London, published 2012; in particular the contributions by G. Affanni, M. Zanon, S. Pizzimenti, K. Afhami and W. Gambke, T. Ornan, deal with different aspects of the use of colour, and of the studies to reconstruct ancient colours.

[3] About the attitude towards 'other' artistic expressions, with special concern for ancient Near Eastern sculpture, see, recently, Evans 2012: 71-72.

[4] It is impossible to mention all the evidences at this regard: see the interesting considerations by Evans 2012: 126-127. It seems meaningful that, while, in general, soft stones of no particular value were used for statues, an exception is represented by a female figure from Nippur, made of translucent green stone, with her face covered by a gold foil (Evans 2012: 128, fig. 44). For the use of composite statues, and sometimes of a peculiar form of polymaterism, implying the use of apparently similar pieces of stone in statuary see Evans 2012: 137-140.

[5] The Eblaic evidence seems particularly interesting if one takes into account the use and location of the artefacts. Whereas the Ur furniture, which is the only comparable evidence, comes from a funerary context, for which it was most likely produced on purpose, the Ebla evidence was part of the palace furniture, it was therefore

decoration were recovered in all the spaces of the Royal Palace G,[6] though it is not always possible to reconstruct their original shape and placement. Some objects, moreover, were certainly unique for their typology and material, and thus even more difficult to interpret. It is possible to maintain, anyhow, that particularly the ceremonial sector of Palace G, including the Court of Audience, the so-called Administrative Quarter, and their annexes, were lavishly decorated with different kinds of fittings where different materials were liberally used, certainly with a strong visual impact (Figure 1).

Among the pieces of furniture which we may consider as specific of north inner Syria, and possibly of Ebla, we may certainly include the wooden furniture (Figure 2). Several charred fragments were found in a room north of the North façade of the Court of Audience, probably belonging to the North Wing of the Central Complex.[7] One table and a chair or throne were identified for certain, which featured several parts with open work decorations of human figures and animal friezes.[8] The wood employed is a fruit tree and ash tree,[9] and the figures were decorated with shell, for the eyes and for further inlays of geometric motifs on the animals' bodies, and for the partitions between registers.[10] A similar use for wood and shell is found in the Trapezoidal Store of the Administrative Quarter, where tablets and precious goods were locked in cupboards, closed by wooden shutters, carved with human heads, probably depicting court officials, whose eyes were made of shell (Figure 3). It is also possible that, in analogy with what happened with other wooden artefacts from the Palace, some parts of the wooden figures were covered with metal leaf, in some cases even with gold (Figure 4).[11] Lastly,

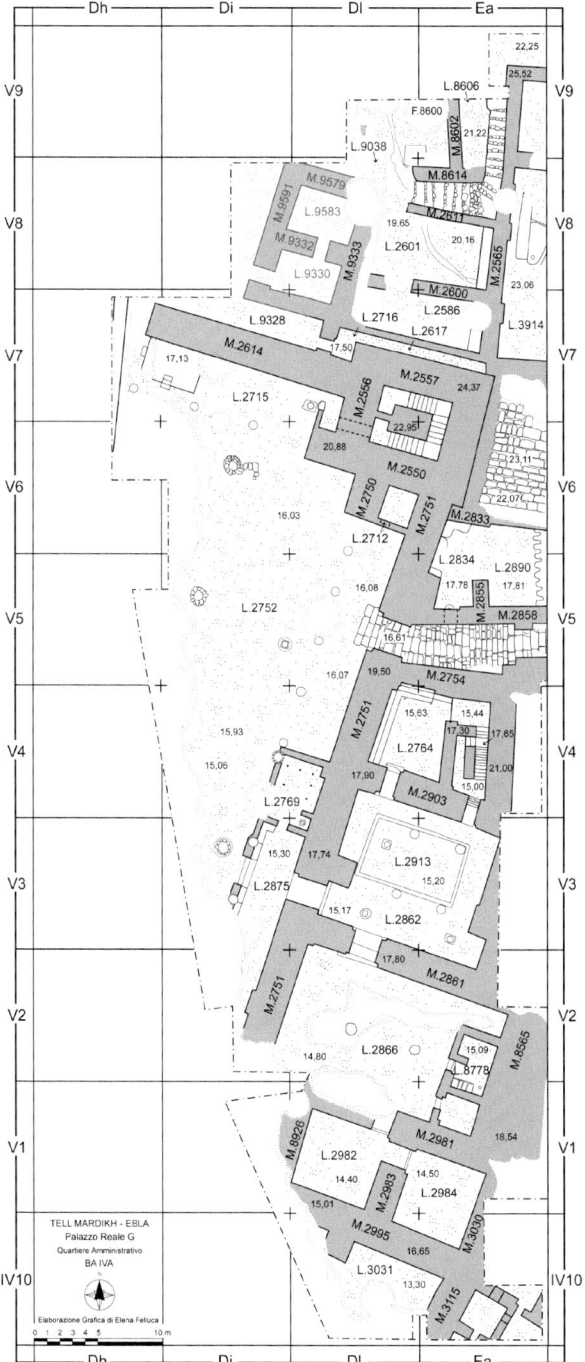

Figure 1. Ebla, schematic plan of the Royal Palace G
© Missione Archeologica Italiana in Siria.

---

very visible, and was certainly part of the language of power in this North Syrian centre.
[6] The exposed quarters of the Royal Palace G certainly belong to the public sectors of the building, the only exception being the stores of the Central Complex located on the south slope of the Acropolis, where, in fact, no fragment of decoration was found. Even the North-West Wing, where service rooms were found, dedicated to food production, was part of the public sector of the Palace, and was related to the suite including rooms L.4424, L.4436 and L.4448, with certainly had a ceremonial function: Matthiae 2017.
[7] Room L.2601, related, and probably communicating with L.2585, where the first 42 tablets of the Royal Archives were found in 1974. The two rooms certainly communicated with the North-West Wing, by means of a staircase, whereas the communication with the lower levels of the Court of Audience is not clear: Matthiae 1995: 77-78.
[8] Matthiae 1985: pls. 37, 43.
[9] Fiorentino and Caracuta 2013: 405, identified a *Pomoidaea* and *Fraxinus* sp. A variety of different wood remains were identified in the Royal Palace G: fir, cedar, olive, plane, poplar. The few preserved fragments of cores of figurines from the Court of Audience were made of olive tree.
[10] The use of triangles to mark the division between registers is a well attested tradition at Ebla, and several stone and shell triangles were found in the Palace debris, which originally belonged to wall panels, or possibly to other pieces of furniture. The use of such triangles in wall panels is proved by the finding of stone elements of this kind in place, together with the figures from the Victory Panel, for which see Matthiae 2017.
[11] The evidence thus far available points to a larger use of gold, as seen in the miniature statue of a human headed bull, in a miniature

wooden planks, with shell inlays in geometric and floral motifs, decorated the steps of the Ceremonial Staircase, in the north-east corner of the Court of Audience, and of the entrance to the Administrative Quarter (Figures 5a, b).[12] The finding of complete or fragmentary locks of lapis lazuli of the same kind as those used for the beards

---

lion-head, and in the remains of the King's Standard (Pinnock 2015), whereas, at present, only Queen Tabur-Damu's figure has silver foil covering the naked parts of her body: Matthiae 2013a: pl. 47a.
[12] Matthiae 1985: pl. 9c; 2010: 79-80, fig. 33; Pinnock 2012: 276.

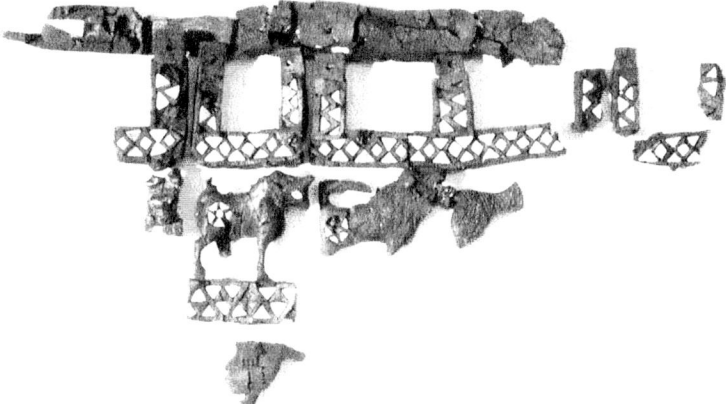

Figure 2. Ebla, arm-rest, probably from a throne, wood and shell, c. 2300 BC, from the Royal Palace G, L.2601
© Missione Archeologica Italiana in Siria.

Figure 3. Ebla, wooden plank with two carved male heads, probably from a cupboard door, wood, from the Royal Palace G, L.2764
© Missione Archeologica Italiana in Siria.

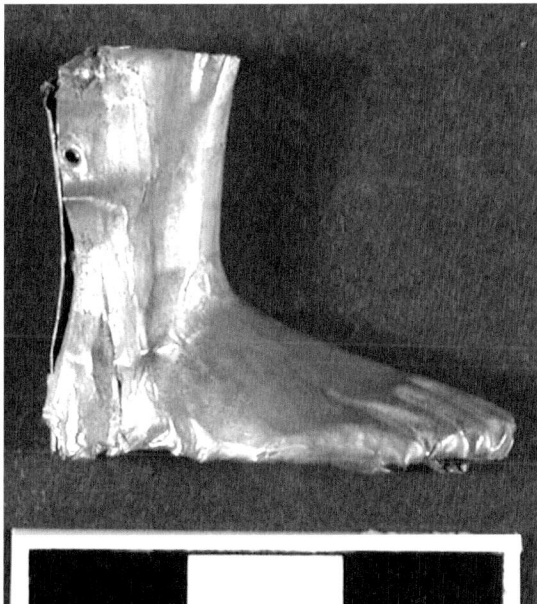

Figure 4. Ebla, revetment of a foot, gold, ca. 2300 BC, Royal Palace G
© Missione Archeologica Italiana in Siria.

of the bulls decorating the musical instruments from the Royal Cemetery of Ur lead to propose that similar instruments were also made at Ebla.[13]

In the Court of Audience, the main decoration included a continuous frieze on wooden planks, c. 10cm high and 2/3cm thick, running all along the North and East façades of the Court, in which stone eyes were inlaid.[14] These eyes, unlike those of statues, were made in one piece only, usually of limestone, and there was no difference between right and left eye, thus allowing for an uninterrupted flow of eyes; statues' eyes were made in three pieces, usually employing limestone and lapis lazuli or steatite, and the difference between right and left eye was clear. It is quite likely that the entrance to the Monumental Stairway was sided by two royal images, either complete figures very close to life-size, or only their heads or busts, whose hair-dresses were made of stone plaques fixed to a wooden core. Of these images only a section of a female hair-dress was preserved (Figure 6), while their reconstruction is based on the presence of similar figures on the sides of the door to the Throne Room.[15] The walls of the Administrative Quarter, which was one of the foci of the ceremonies of kingship, were decorated with inlaid panels of two main types:[16] the first one was closer to Early Dynastic Mesopotamian models, and included small wooden panels, with figures made mainly in limestone, but also of shell, depicting war scenes, or contest scenes with wild and domestic animals (Figures 7a, b).[17] The inlays were made with the silhouette technique, with incised details on a flat plaque, but smaller inlays might be used in order to represent the ropes tying the hands of a prisoner (Figure 8), horses'

---

[13] As already maintained (Pinnock 1984: 26) the locks found at Ebla were locally made, as they are slightly different from those of Ur, which are sometimes quite long tufts, ending in a lock, whereas the Eblaic pieces are always very short. The same holds true for a small lump of bitumen, where small lozenges of lapis lazuli were applied (Pinnock 1985: 135-136, fig. 5), whose only comparison may be found in the sledge-chariot from dame Pu-Abi's tomb at Ur: Woolley 1934: pls. 122-126, U.10438.
[14] Matthiae 2008: 53; Pinnock 2012: 275.
[15] The fragment was found on the last step of the Monumental Staircase: Matthiae 2013b: 425-426, 429-430, pl. 54a.
[16] Matthiae 2010: 164-167.
[17] Matthiae 1985: pl. 47.

Figure 5a. Ebla, the steps of the Ceremonial Staircase, ca. 2300 BC, Royal Palace G © Missione Archeologica Italiana in Siria.

Figure 5b. Ebla, entrance to the Administrative Quarter, with limestone threshold and wood and shell decoration, ca. 2300 BC, Royal Palace G, L.2875 © Missione Archeologica Italiana in Siria.

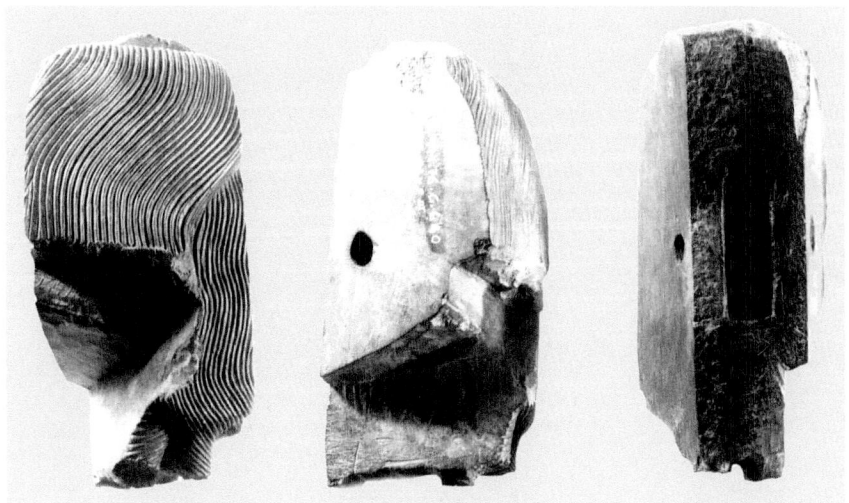

Figure 6. Ebla, three views of a segment of female hair-dress, steatite, ca. 2300 BC, Royal Palace G, L.2752, on the last step of the Monumental Stairway © Missione Archeologica Italiana in Siria.

bridles, coloured spots in animals furs, or to fill the gaps between figures: most of these inlays are lost, but a few fragments of bridles and of background fragments, mainly of lapis lazuli, are still preserved (Figure 9).[18]

The most typical Eblaic panel is larger in size, and represents processions of palace officials leading towards a royal figure: the officials are in profile, whereas the king is front facing and slightly larger (Figures 10a, b). In these panels, the wooden base was worked in low relief, in order to represent the human bodies: the modelled naked parts of the bodies were covered with gold leaf, as is proved by the finding of several fragments of gold leaf representing hands or legs or feet (Figure 4); the skirts, cloaks and belts were made of limestone, in separate pieces; the officials' hair-dresses were of lapis lazuli, and the king always wore his typical turban, made of limestone. Stone rosettes, found in a large number, were probably, at least in part, the decoration of the borders of the panels (Figure 11). The stone components of the panels were held in place by pegs coming out from the wood bases, and lodged in holes on the back of the stone pieces, whereas the gold leaf was nailed to the wooden background, by means of minute gold nails, some of which were still attached to the metal leaves, and several were found loose in

---

[18] Similar elements are also present at Mari, in shell inlays: Margueron 2004: 293, fig. 28, 14, as separate pieces, and 17-18, carved in one piece with the head of the animal.

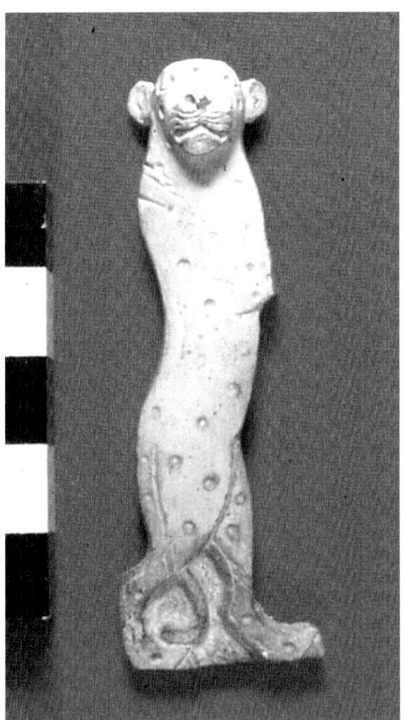

Figure 7a. Ebla, figure of standing leopard from a wall inlaid panel, limestone with lost inlays of different materials, ca. 2300, Royal Palace G, L.2913 © Missione Archeologica Italiana in Siria.

Figure 7b. Ebla, figure of passing human-headed bull from an inlaid wall panel, limestone with lost inlay for the beard, ca. 2300 BC, Royal Palace G, L.2913 © Missione Archeologica Italiana in Siria.

Figure 9. Ebla, fragments of bridles from wall inlays, lapis lazuli, ca. 2300 BC, Royal Palace G © Missione Archeologica Italiana in Siria.

Figure 10a. Ebla, reconstruction of a procession of officials from an inlaid wall panel, limestone and lapis lazuli, ca. 2300 BC, Royal Palace G © Missione Archeologica Italiana in Siria.

Figure 8. Ebla, figure of standing prisoner, limestone with lost inlays for the rope holding his arms and for the background, ca. 2300 BC, Royal Palace G, L.2913 © Missione Archeologica Italiana in Siria.

Figure 10b. Ebla, reconstruction of three front facing kings' figures from inlaid wall panels, limestone, ca. 2300 BC, Royal Palace G © Missione Archeologica Italiana in Siria.

Figure 11. Ebla, three rosettes from the decoration of inlaid wall panels, limestone, ca. 2300 BC, Royal Palace G © Missione Archeologica Italiana in Siria.

Figure 12. Ebla, three views of a composite female hair-dress, steatite, ca. 2300 BC, Royal Palace G, L.2862, at one side of the entrance to the Throne Room L.2866 © Missione Archeologica Italiana in Siria.

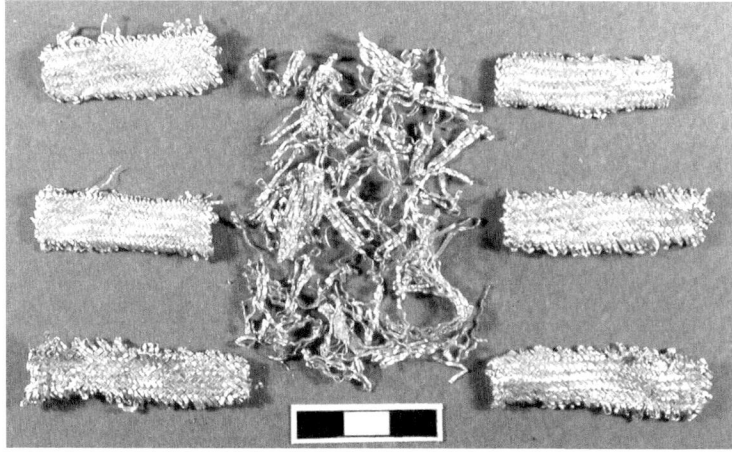

Figure 13. Ebla, fragments of a textile, or of the decoration of a textile, gold, ca. 2300 BC, Royal Palace G. L.8778 © Missione Archeologica Italiana in Siria.

the soil.[19] As mentioned before, a royal couple in the round was placed at both sides of the entrance to the Throne Room. Only the hair-dresses were found: these were made of separate plaques of steatite, originally mounted on a wooden base and probably completed after mounting, as the excellent workmanship of some of the locks, represented over two adjoining plaques, leads to believe (Figure 12). It is quite likely that the faces were covered with precious metals, silver or gold, while it is not possible to understand if the statues originally included the full figure. In this case, anyhow, certainly they did not feature stone clothing, as no element large enough was found in the court of the Administrative Quarter; on the other hand, it is possible that some precious textile, kept in the two accessory rooms inside the Throne Room, was used to cover the eventual wooden statues (Figure 13).[20]

Other, most peculiar objects are those we identified as standards: in our reconstruction, the Eblaic officials during ceremonies carried a cane or a staff, on which a small figure in the round was applied, probably pointing to their status or role.[21] According to this hypothesis, all, or most of, the miniature figures in the round, several complete or fragmentary specimens of which were found in Palace G, were precisely the decorations of these standards. All these figures have holes in their bases, pointing to the fact that they had to be fixed to some kind of support. To this typology certainly belong the figurines of squatting bulls or rams, whose fur was sometimes decorated with other stone inlays,[22] but possibly also a human-headed bull, made of wood, gold and steatite (Figure 14).

Several fragments of human figures in the round were also recovered, together with a complete image of a veiled woman, of steatite, limestone and jasper (Figure 15),[23] which might have belonged to standards where

---

[19] Matthiae, Pinnock and Scandone Matthiae 1995: 326-328, nos. 114-120, for several examples of gold leaf, some featuring the holes for the passage of the nails, and some miniature gold nails.

[20] From L.8778, a small room built inside the Throne Room of the Palace: Matthiae 2013c: 480-481, pl. 70a.
[21] Pinnock 2015: 5.
[22] Matthiae 2013d: figs 58a-c.
[23] A woman wearing the same kind of dress is represented in a relief plaque from Ebla, found in the levels belonging to Building G5, the

Ebla (Figure 16),[24] and the probable reconstruction of the en's standard,[25] on the other hand, lead to propose that the human figures rather belonged to composite standards, belonging to the topmost levels of the palace hierarchy. To this typology I think also a small wooden head of a lion, covered with gold leaf and mounted on a mobile support should be related (Figure 17).[26] The two composite standards thus far reconstructed, with a different degree of certainty, were found one – the *maliktum*'s standard – in a room north of the Court of Audience, and in communication with it,[27] whereas the en's standard, which was much

Figure 14. Ebla, miniature figure of human-headed bull, possibly from a standard, gold and steatite, ca. 2300 BC, Royal Palace G, L.2764
© Missione Archeologica Italiana in Siria.

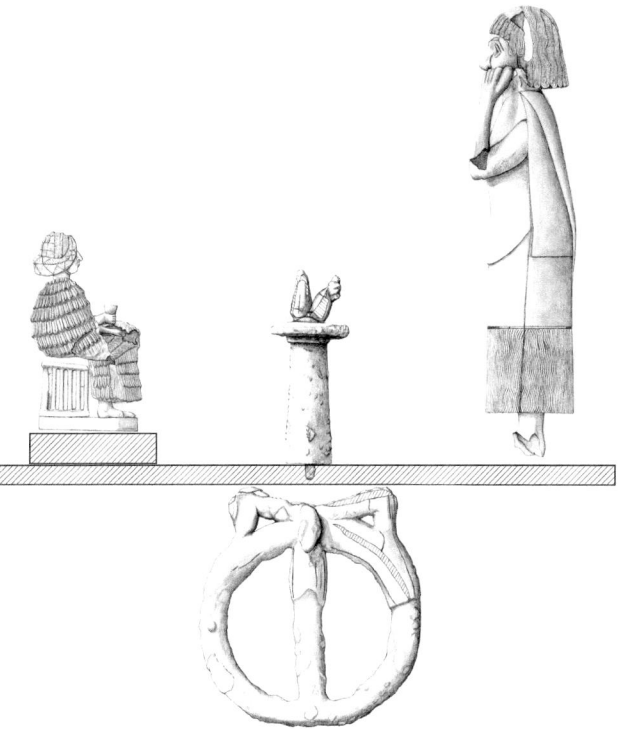

Figure 16. Ebla, reconstructive drawing of the *maliktum*'s standard, wood, steatite, limestone, gold, silver and jasper, ca. 2300 BC, Royal Palace G, L.9330
© Missione Archeologica Italiana in Siria.

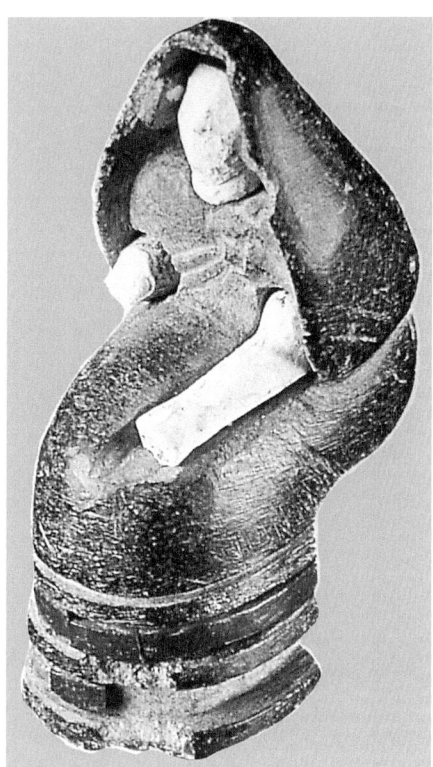

Figure 15. Ebla, miniature figure of veiled woman probably from a standard, limestone, steatite and jasper, ca. 2300 BC, Royal Palace G, L.3600
© Missione Archeologica Italiana in Siria.

they were the only element, like the squatting animals. The recent discovery of the *maliktum*'s standard, probably belonging to Tabur-Damu, last queen of

---

[24] Matthiae 2013a.
[25] Pinnock 2015: 10-20, fig. 19.
[26] In this case, apparently the figure did not have a solid core, because the wooden support is composed of two studs, one connecting the two ears, and the other inserted in the first at a right angle, in order to connect the head to the rest of the figure, or to the eventual base, and to leave the head free to move. The head might recall the lion-headed eagle from the so-called Treasure of Ur from Mari (Parrot 1968: M. 4405, 23, fig. 16, pls. IX-X), but with some important stylistic and working differences. The Eblaic head is completely smooth, whereas the Mari piece features chevron motifs on the muzzle and behind the ears; in the Eblaic piece the ears are large and slightly oblique, whereas in the Mari piece the ears are small, straight and narrower at the bottom. As regards technique, the Mari head has a solid bitumen core, and it therefore was fixed to the lapis lazuli body.
[27] L.9583: the pieces were scattered on the floor at a close distance one from the other, thus making the hypothesis that they belonged to one piece quite probable; at a short distance, also the remains of the wooden pole supporting the standard, and of its bronze covering were also found in a very bad state of preservation: Matthiae 2013a: 459, figs. 30b, 50b.

predecessor of the Royal Palace G (Dolce 2008; Vacca 2015: 4; Pinnock in press a and b), and a similar character is depicted on a sealing from Mari, from a tomb: Margueron 2004: 297, fig. 285.3-4.

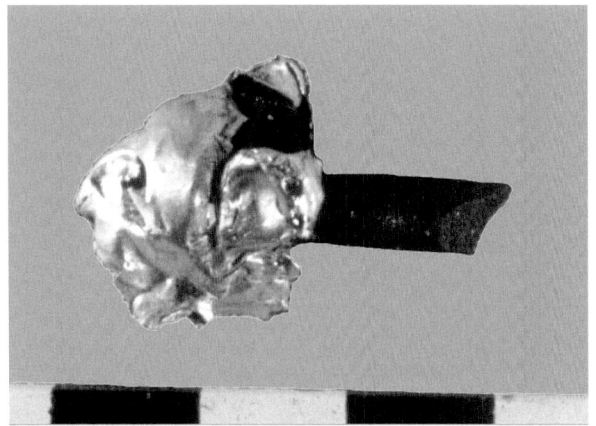

Figure 17. Ebla, miniature head of lion on a mobile support, wood and gold, ca. 2300 BC, Royal Palace G, L.2984 © Missione Archeologica Italiana in Siria.

of jasper decorates the partition between the smooth textile and the fringe. The steatite hair-dress is made of five preserved elements, but two parts at least are missing: one section is the back part of the hair-dress (Figure 19a), the fringe includes two elements and one lock is on the side of the face. One more lock should probably be reconstructed on the other side, while the back part of the hair-dress is separated from the fringe by a thick silver band. Also the hair-dress is accurately worked at the bottom (Figure 19b), which leads to infer that the object had to be seen from below, and confirms its interpretation as a standard. Eyes and eyebrows were inlaid, but only the limestone socket of the left eye is preserved. So, this image included 13 or more individual pieces (Figure 20).

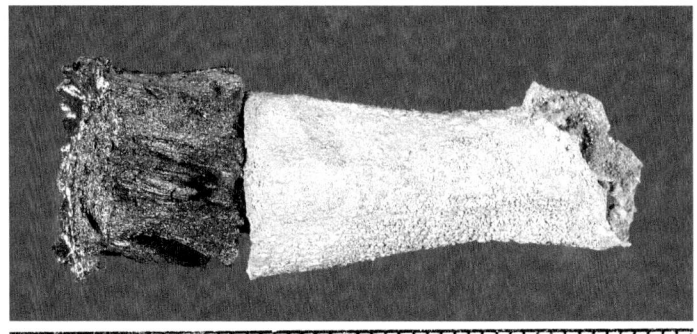

Figure 18a. Ebla, wooden core of Tabur-Damu's figure from the *maliktum*'s standard, showing the complete modelling of the face, ca. 2300 BC, Royal Palace G, L.9330 © Missione Archeologica Italiana in Siria.

Figure 18b. Ebla, arm of Tabur-Damu's figure from the *maliktum*'s standard, showing the modelled wooden core and the thick silver coating, ca. 2300 BC, Royal Palace G, L.9330 © Missione Archeologica Italiana in Siria.

more fragmentary, was kept in one of the rooms south of the Throne Room.[28] Both certainly had a base of wood, for the support and for the core of the figures. The composition of these objects is quite complex. As regards the *maliktum*'s standard, the largest figure, identified as Tabur-Damu, last queen of Ebla, has her face, bust, arms and legs completely modelled in wood (Figure 18a), and covered with a thick silver leaf (Figure 18b). Her dress is a cloak made of one block of steatite, carved inside, in order to fit over the wooden core, and accurately worked in the lower part, where the end of the long fringe is; a thin band

The smaller, seated figure, identified with the funerary statue of Dusigu, Tabur-Damu's mother-in-law and predecessor, is made of different materials and with a different technique. The wooden base was not carved, the naked parts of the body were made of limestone, and were fixed on the base; the flounced dress included, according to the Mari style,[29] a cape and the dress proper, which are made of one piece of gold foil, worked with the *repoussé* technique (Figure 21), and cut in

---

[28] There is a correlation between the Court of Audience and its adjoining rooms, and the Throne Room and its adjoining rooms, certainly related to the kind of ceremonies which took place in both: Pinnock 2012: 277-279.

[29] The particular composition of this attire, very close to the dress worn by the priestesses' statues from Mari, and the connection between a 'dress in Mari style' and funerary cults in the Rituals from Ebla (Fronzaroli 1993: passim), led me to propose that this specific *kaunakés* with cape were precisely the dress in the Mari style of the Eblaic text: Pinnock 2016; in press c. For closely comparable specimens: Parrot 1967: pl. LVII, M. 2788, where the figure has long loose hair and Margueron 2004: 281, fig. 266, a priestess wearing the polos.

Figure 19a. Ebla, back part of the hair-dress of Tabur-Damu's figure from the *maliktum*'s standard, steatite, ca. 2300 BC, Royal Palace G, L.9330
© Missione Archeologica Italiana in Siria.

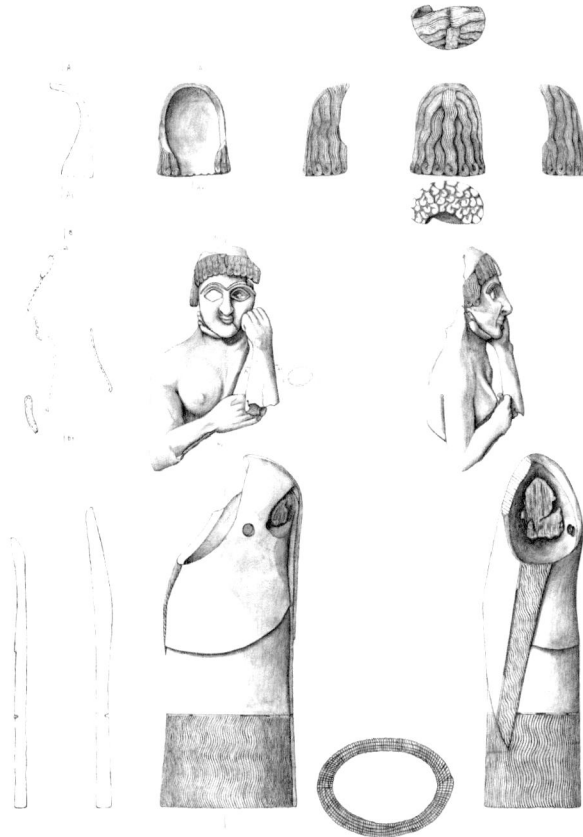

Figure 20. Ebla, drawing of the separate pieces composing Tabur-Damu's figure from the *maliktum*'s standard, wood, steatite, silver and jasper, ca. 2300 BC, Royal Palace G, L.9330
© Missione Archeologica Italiana in Siria.

Figure 19b. Ebla, lower edge of the hair-dress of Tabur-Damu's figure from the *maliktum*'s standard, steatite, ca. 2300 BC, Royal Palace G, L.9330
© Missione Archeologica Italiana in Siria.

Figure 21. Ebla, dress of Dusigu's figure from the *maliktum*'s standard, gold, ca. 2300 BC, Royal Palace G, L.9330
© Missione Archeologica Italiana in Siria.

order to host the limestone arms. The queen wore an elaborate steatite head-dress, probably representing a cloth turban. As happens with Tabur-Damu, eyes and eyebrows are lost, with the exception of the steatite left eye's socket. The queen holds in her right hand a tiny jasper beaker, and is seated on a square limestone stool, whose fluted sides were inlaid with jasper, and were covered with a gold foil, cut in order to leave the jasper inlays visible. In order to further stress the fact that this image represented a statue, and not a living person, it was placed on a rectangular steatite base. Overall, this tiny figure is composed of 13 individual

Composite Artefacts in the Ancient Near East

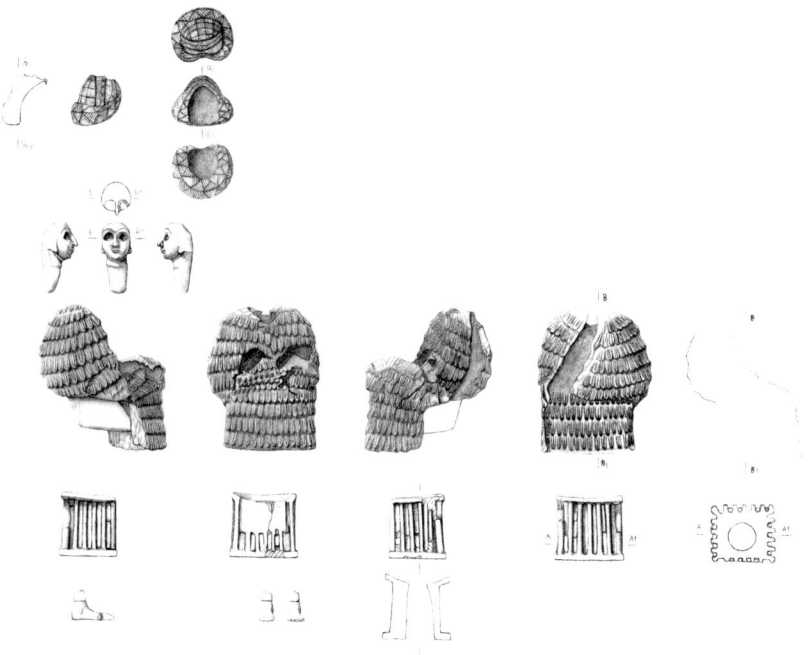

Figure 22. Ebla, drawing of the separate pieces composing Dusigu's figure from the *maliktum*'s standard, wood, steatite, gold, marble and jasper, ca. 2300 BC, Royal Palace G, L.9330 © Missione Archeologica Italiana in Siria.

pieces (Figure 22).[30] The en's standard is much more fragmentary and its reconstruction more speculative. Among the scattered fragments of palace fittings and precious goods kept in the square store behind the Throne Room of the Administrative Quarter, a beautiful steatite hair-dress and the remains of two arms covered in gold leaf were immediately attributed to a figure of bull-man, and consequently also an almost complete horn in gold foil, perfectly adapted to a hole opening into the hair-dress was also attributed to this figure.

In a later analysis of the complex of the evidence from this room, I proposed to reconstruct another three-figures standard, pivoting on the previously reconstructed bull-man. In my analysis, this second standard probably belonged to the king, and reproduced a kind of royal emblem, also attested in the Eblaic cylinder seals, with the king and queen on both sides of the bull-man.[31] The basic structure of the standard was quite likely similar to that of the *maliktum*'s standard, with a wooden core for the pole and figures. The preserved elements are the steatite hair-dresses of the three personages, the golden foil covering the bull-man's arms and horn,

one foot and one hand belonging to one or the other of the human characters, a fragment of the queen's shell diadem, a fragment of the king's limestone beard and the bull-man's red stone tail (Figure 23). The two standards were comparable in general size, as Tabur-Damu's figure was of the same height as the three figures of the en's standard. In this reconstruction, it seems possible to maintain that the king and queen, and perhaps also some official of a very high rank, like the vizier, had composite standards, quite likely carried by servants accompanying them, whereas the other court members had simpler standards, made of one figure only, which they could carry by themselves. It appears quite evident that the sectors of the Royal Palace G thus far brought to light were meant to be a real theatrical scenery for the ceremonies taking place in all of them, by means of the combined effect of the multi-coloured decoration of the walls and of the presence of refined inlaid furniture. Its occupants, on the other hand, had to enact and embody a specific role, which was also manifest in their clothing, ornaments and specific paraphernalia.

As I have maintained elsewhere,[32] there was a colour code in dressing, which is also partially reflected in statuary.[33] On the other hand, it is quite difficult to elaborate on the eventual aesthetic meaning of these composite objects. This theme is complex for two main reasons: the first one I have recalled at the beginning, namely that we have a partial knowledge of the real final aspect of the objects whose fragments we take into account; the second, and more serious one, is that we ignore the taste and aims of those who had requested these artefacts. The archival texts of Ebla – mainly administrative texts – never mention the wish to meet a specific taste need. At Ebla there is no text of the type of the LUGAL UD ME-LÁM-bi NIR-GÁL, from which we may in part infer the Sumerian consideration for the different materials.[34] On the other hand, we might imagine some similarity

---

[30] The fragments found include (1) the hairdress, (2) the face, (3) parts of the wood base, (4) the right arm; (5) the left arm holding a (6) cup, (7) the *kaunakés*, (8–9) the feet, (10–12) the stool with its gold foil covering and several fragments of jasper inlays inserted in the grooves of its faces, (13) the base. In the drawing presented here the hands are missing.
[31] The figurative patterns of cylinder seals also include, following the Early Dynastic south Mesopotamian tradition, bulls attacked by lions, which are protected by a goddess and the bull-man, aided by the king and queen: Pinnock 2013.

[32] Pinnock 2015: 279.
[33] It is most likely that the different materials used in order to represent clothing were not casually chosen, but rather aimed at reproducing the real colours of the original dresses.
[34] For a classical edition of this text see van Dijk 1983: in particular vol. I on pp. 19-25, 44-47, 104-136.

Figure 23. Ebla, proposal of reconstruction of the en's standard, steatite, gold, limestone, red stone and shell, ca. 2300 BC, Royal Palace G, L.2982 © Missione Archeologica Italiana in Siria.

with what happens at Mari, a few centuries later, where apparently craftsmen were praised for their technical capacities. The verbs referring to the manufacture of goods are *dummuqum* 'to improve', *bunnûm* 'to beautify', *idûm* 'to know', *le'ûm* 'to be capable', *nukkulum* 'to execute in a refined manner', *šuklulum* and *quttum* 'to bring to a (perfect) end'.[35] Craftsmen are defined *mudûm* and *taklum* 'competent and reliable, experienced',[36] but they are usually classified according the material they were used to work, as if material was more important than the way of handling it.[37]

If this set of values was true also for the earlier Ebla court, undoubtedly the Eblaic craftsmen were capable to meet the needs of their purchasers at the highest level: they could profit of the abundance of wood of different qualities present in the nearby mountains, while precious imported materials were used in the creation of composite objects where technical skills of different kinds were employed with great mastery, from the accurate carving of the invisible wooden bases of inlays and miniature statues, to the delicate assemblage of the different parts of an individual object, using dovetail joints for the wooden parts, bitumen and other glues for stone, and diminutive gold nails to fix the gold foil. In the *maliktum*'s standard we can also probably appreciate a form of realism, with the attempt at reproducing a statue, with the naked parts of the body made of limestone, and a living body, with the naked parts of the body covered with silver foil, which being originally white and brilliant was probably meant to reproduce the white, perfect and glowing skin of a paradigm of beautiful young woman.[38]

The general picture we gain from the observation of the architectural setting of the Royal Palace G and of its decorative pattern, as far as we can reconstruct it, in combination with the hypothesis we may propose about the attires and ornaments of its occupants, is one of an accomplished, manifold ensemble of manifestations of kingship. In a monumental setting, the court élite enacted a ritual, hieratic staging, reproduced on the walls in the inlaid panels, where colourful clothing and refined artefacts contributed in creating a powerful image of the court and its members in ways which more and more appear specific to the North Syrian, or rather Eblaic culture.

References

Affanni, G. 2012. New Light (and Colour) on the Arslan Tash Ivories: Studying 1st Millennium BC Ivories. In Matthews and Curtis (eds): 193-208.

Afhami, K. and W. Gambke 2012. Colour and Light in the Architecture of Persepolis. In Matthews and Curtis (eds): 335-349.

van Dijk, J. 1983. LUGAL UD ME-LÁM-bi NIR-GÁL. *Le récit épique et didactique des travaux de Ninurta, du déluge et de la nouvelle création*. Leiden: Brill.

Dolce, R. 2008. Ebla before the Achievement of the G Palace Culture: an Evaluation of the Early Syrian Archaic Period. In H. Kühne, R.M. Czichon and F.J. Krepner (eds), *Proceedings of the 4th International Congress on the Archaeology of the Ancient Near East, 29 March-3 April 2004*, vol. 1: 65-80. Wiesbaden: Harrassowitz.

Evans, J.M. 2012. *The Lives of Sumerian Sculpture. An Archaeology of the Early Dynastic Temple*. Cambridge: Cambridge University Press.

Fiorentino, G. and V. Caracuta 2013. Forests Near and Far. An Anthracological Perspective on Ebla. In Matthiae and Marchetti (eds) 2013: 403-412.

---

[35] Sasson 1990: 22.
[36] Sasson 1990: 22.
[37] Sasson 1990: 23. The Akkadian term *mār ummênim* usually refers to a number of specialistic capacities, from textile working to wood or leather working, to singing and brewing. It apparently refers to technical, rather than artistic capacities. For an illuminating analysis of the consideration of art works in the Neo-Assyrian period see Matthiae 2014, and for important considerations on the meaning of the use of gold Winter 2012.

[38] Tabur-Damu's counterpart, on the other hand, has a curving nose and double chin, with a general aspect of more mature age, as representation of reality – Dusigu was dead when the standard was made – but also as a mark of different role and prestige between the two women.

Fronzaroli, P. 1993. *Archivi Reali di Ebla. Testi XI. Testi rituali della regalità (Archivio L. 2769)*. Roma: 'Sapienza' Università di Roma.

Margueron, J.-Cl. 2004. *Mari. Métropole de l'Euphrate au IIIe et au début du IIe millénaire av. J.-C.* Paris: Picard.

Matthews, R. and J. Curtis (eds) 2012. *Proceedings of the 7th International Congress on the Archaeology of the Ancient Near East, London, 12 April – 16 April 2010, vol. 2*. Wiesbaden: Harrassowitz.

Matthiae, P. 1985. *I tesori di Ebla*. Roma-Bari: Laterza.

Matthiae, P. 1995. *Ebla. Un impero ritrovato. Dai primi scavi alle ultime scoperte*, 3rd edition. Torino: Einaudi.

Matthiae, P. 2008. *Gli Archivi Reali di Ebla. La scoperta, i testi, il significato*. Milano: Mondadori.

Matthiae, P. 2010. *Ebla. La città del trono. Archeologia e storia*. Torino: Einaudi.

Matthiae, P. 2013a. The Standard of the *maliktum* of Ebla in the Royal Archives period. In Matthiae, *Studies on the Archaeology of Ebla 1980-2010* (ed. F. Pinnock): 455-477. Wiesbaden: Harrassowitz.

Matthiae, P. 2013b. Some Fragments of Early Syrian Sculpture from Royal Palace G of Tell Mardikh-Ebla. In P. Matthiae, *Studies on the Archaeology of Ebla 1980-2010* (ed. F. Pinnock): 419-438. Wiesbaden: Harrassowitz.

Matthiae, P. 2013c. The Seal of Ushra-Samu, Official of Ebla, and Ishkhara's Iconography. In P. Matthiae, *Studies on the Archaeology of Ebla 1980-2010* (ed. F. Pinnock): 479-493. Wiesbaden: Harrassowitz.

Matthiae, P. 2013d. About the Style of a Miniature Animal Sculpture from the Royal Palace G of Ebla. In P. Matthiae, *Studies on the Archaeology of Ebla 1980-2010* (ed. F. Pinnock) (Wiesbaden 2013) 439-454.

Matthiae, P. 2014. Fire and Arts. Some Reflections about the Consideration of Art in Assyria. In P. Bielinski, M. Gawlikowski and R. Kolinski (eds), *Proceedings of the 8th International Congress on the Archaeology of the Ancient Near East, Warsaw, 30 April – 4 May 2012, vol. 3*: 93-121. Wiesbaden: Harrassowitz.

Matthiae, P. 2017. The Victory Panel of Early Syrian Ebla in its Historical Context: Finding, Structure, Dating, *Studia Eblaitica* 3: 33-83.

Matthiae, P. and N. Marchetti (eds) 2013. *Ebla and its Landscape. Early State Formation in the Ancient Near East*. Walnut Creek (CA): Left Coast Press.

Matthiae, P., F. Pinnock and G. Scandone Matthiae (eds) 1995. *Ebla. Alle origini della civiltà urbana. Trent'anni di scavi in Siria dell'Università di Roma 'La Sapienza'*. Milano: Electa.

Ornan, T. 2012. The Role of Gold in Royal Representation: the Case of a Bronze Statue from Hazor. In Matthews and Curtis (eds): 445-458.

Parrot, A. 1967. *Mission archéologique de Mari, III. Les temples d'Ishtarat et de Ninni-zaza*. Paris: Paul Geuthner.

Parrot, A. 1968. *Mission archéologique de Mari, IV. Le 'Trésor' d'Ur*. Paris: Paul Geuthner.

Pinnock, F. 1984. Trade at Ebla, *Bulletin of the Society for Mesopotamian Studies* 7: 19-31.

Pinnock, F. 1985. Einige Erwägungen zum Handel von Ebla, *Das Altertum* 31: 133-140.

Pinnock, F. 2012. Colours and Light in the Royal Palace G of Early Syrian Ebla. In Matthews and Curtis (eds): 271-286.

Pinnock, F. 2013. Palace vs Common Glyptic Style in Early Syrian Ebla and its Territory. In Matthiae and Marchetti (eds): 66-72.

Pinnock, F. 2015. The King's Standard from Ebla Palace G, *Journal of Cuneiform Studies* 67: 3-22.

Pinnock, F. 2016. Dame di corte a Ebla: aspetto, ruoli e funzioni delle donne in una grande corte protosiriana. In M.G. Biga, L. Mori and F. Pinnock (eds), *Donne d'Oriente. Voci e volti di donne nel Mediterraneo orientale*, Henoch 38/2. Brescia: Morcelliana Edizioni.

Pinnock, F. in press a. Ebla in the IIIrd Millennium BC. Architecture and Urban Planning. In St. Bourke and M. Kennedy (eds), *Proceedings of the Workshop 'The Settlement Landscape of the Orontes Valley in the Fourth Through Second Millennia BCE'*, Studia Chaburiensia.

Pinnock, F. in press b. Ebla in the Mid- to Late III Millennium BC: Architecture and Chronology. In S. Richard (ed.), *New Horizons in the Study of the Early Bronze III and Early Bronze IV in the Levant*. Winona Lake: Eisenbrauns.

Pinnock, F. in press c. About the Attire of High Rank Ladies in Early and Old Syrian Ebla, *Studia Eblaitica*.

Pizzimenti, S. 2012. Colours in Late Bronze Age Mesopotamia. Some Hints on Wall Paintings from Nuzi, Kar-Tukulti-Ninurta and Dur Kurigalzu. In Matthews and Curtis (eds): 303-317.

Sasson, J. 1990. Artisans ... Artists: Documentary Perspectives from Mari. In A.C. Gunter (ed.), *Investigating Artistic Environments in the Ancient Near East*: 21-27. Washington (DC): Smithsonian Institution.

Vacca, A. 2015. Before the Royal Palace G. The Stratigraphic and Pottery Sequence of the West Unit of the Central Complex: the Building G5, *Studia Eblaitica* 1: 1-32.

Winter, I.J. 2012. Gold! Divine Light and Lustre in Ancient Mesopotamia. In Matthews and Curtis (eds): 153-171.

Woolley, C.L. 1934. *Ur Excavations II. The Royal Cemetery. A Report on the Predynastic and Sargonid Graves Excavated between 1926 and 1931*. Oxford: University Press.

Zanon, M. 2012. The Symbolism of Colours in Mesopotamia and the Importance of Light. In Matthews and Curtis (eds): 221-243.

Near Eastern Materials, Near Eastern Techniques, Near Eastern Inspiration: Colourful Jewellery from Prehistoric, Protohistoric and Archaic Cyprus

Anna Paule
Independent researcher from Linz
annapaule.archaeo@hotmail.com

Abstract

The goldsmith's art from Prehistoric, Protohistoric and Archaic Cyprus is exemplified by funerary jewellery made of thin gold sheet, typically bearing a *repoussé* pattern. The aim of this article, however, is to survey the more elaborate, colourful jewellery from ancient Cyprus, whose components were assembled from a large variety of materials, such as metals, semi-precious stones, and vitreous materials as well as ivory or related animal materials. This production method was mainly used for manufacturing dress pins, finger rings, and necklaces which were made by combining two or more distinct components. The materials, manufacturing techniques and symbols used in jewellery making clearly reflect Near Eastern goldsmithing traditions, indicating that these jewellery items were partly, if not entirely, imported from the Near East. The present article considers not only the striking parallels between Cypriot and Near Eastern jewellery, but also further research issues, such as the developments of the cloisonné technique and enamelling for jewellery making, which await further clarification through future research.

Keywords: colourful jewellery, Prehistoric/Protohistoric Cyprus, Cypro-Archaic period, faience/ivory pinheads, *cloisonné* technique, Near Eastern jewellery, goldsmithing techniques

Following L. and P. Åström's systematic classification of Late Cypriot jewellery types, including important remarks on Near Eastern forerunner models, the joint study of Cypriot and Near Eastern jewellery has attracted surprisingly little interest.[1] This also applies to colourful jewellery from Cyprus that displays a great variety of Near Eastern materials (semiprecious stones, faience) and goldsmithing techniques (especially granulation and inlay). This type of jewellery was mainly used as dress fastening and for adorning fingers or necks.

Jewellery can be divided into two classes based on the techniques and materials used in its manufacture: (1) composite jewellery items consisting of two or more distinct materials; and (2) gold jewellery with colourful designs. Here, the focus is primarily on the outstanding pieces which go beyond the usual repertoire of Cypriot jewellery but find striking parallels in the ancient Near East.[2] Good examples include a series of dress pins and finger rings from Enkomi and Kouklia-Evreti, and also two colliers that were discovered at Enkomi and Arsos during the early days of Cypriot archaeology. This article, however, aims to go beyond merely listing colourful jewellery in order to provide a more qualitative analysis. More precisely, the aim of this article is twofold. Firstly, to provide more detailed information about Near Eastern forerunner models of colourful jewellery found at Cyprus. Secondly, to encourage further research by identifying gaps and inconsistencies in the existing research. In addition, special emphasis is given to manufacturing techniques, such as the first application of the *cloisonné* technique or enamelling for jewellery making, research issues which remain an open chapter of technological history. Additionally, a separate chapter is devoted to the colours and materials which were used in jewellery making, offering a further route to the understanding of the history and development of jewellery of ancient Cyprus. However, due to the limited scope of this article, the investigation is based on selected examples of jewellery from different periods and cannot claim to be exhaustive.

Composite Jewellery Items I. Dress Pins combining Gold and Faience

The starting point of this investigation is some colourful jewellery from Late Bronze Age tombs at Enkomi. The

---

[1] Åström and Åström 1972: 499–511, 567–583. More recent extensive surveys on Cypriot jewellery, which include detailed comments on Near Eastern forerunner models, are the unpublished PhD theses of Goring 1983 on Late Cypriot jewellery, and Kontomichali 2002 on Cypro-Geometric jewellery.
[2] Jewellery items of little distinctive character, such as single beads or components of jewellery that can be assigned only generally to the ancient Near East, were deliberately excluded from this study.

first examples are a series of four loop pins or toggle pins with an attached loop[3] which were found in British Tombs (hereafter abbreviated as Br. T.) 19 and 92, as well as in Swedish Tombs (hereafter abbreviated as Sw. T.) 17 at Enkomi. This type of Cypriot dress pin classed as a composite jewellery item is assembled from three different parts: (1) the shaft of the gold pin, whose elongated upper part is ornamented with gold wire or discs; (2) an additional loop made of gold wire that is strung halfway along its length; and (3) a faience bead that is attached to the top of the pin. In addition, two further loop pins were unearthed in Br. T. 93 and 67 at Enkomi that differ slightly from the aforementioned pins: the first pinhead is missing (Br. T. 93), and in the second, the loop pin was made entirely of gold (Br. T. 67).

A closer inspection of Cypriot loop pins reveals further particularities. Four specimens – Br. T. 19, 92–93 and Sw. T. 17 at Enkomi – are characterised by a shaft whose upper part seems to be fashioned into a so-called double or multiple loop-in-loop chain.[4] Ancient loop-in-loop chains are made of intricately woven gold wire. This manufacturing technique, however, is very complex and time-consuming because these chains are assembled from individual wire loops which were first soldered and then folded and fitted together. In the Eastern Mediterranean, gold loop-in-loop chains appear as early as 2500 BC (in the jewellery of Mochlos), but are especially well known from ancient Egypt; occasionally, they are encountered in Mycenaean Greece.[5] Loop pins from Cyprus may be further examples of this manufacturing technique, but E. Goring, who studied them within the scope of her PhD thesis, indicates that the herringbone pattern along the pins was made using pairs of plaited wire that were soldered together.[6] Additionally, a blue or white faience bead was fastened to the top of these Cypriot loop pins (Br. T. 19 and 92, LC IA/IB–IIC. LCIIA?).[7] In one case (Sw. T. 17/II, LC II), a fluted gold bead of similar size joins the blue faience bead to the shaft of the pin.[8] In addition to these pins, another specimen that may be of the same type is known from Enkomi, but its former bead or pinhead has been lost (Br. T. 93, LCIA–IIC. LCIII?).[9] Apart from this specimen, most of the faience pinheads are well preserved and can be classified as melon beads; only the bead attached to the loop pin from Br. T. 92 is less well preserved and can no longer be classified as such (Figure 1).

The custom of using colourful beads as heads of composite pins dates back to the 3rd millennium BC, when pinheads made from semiprecious stones, such as lapis lazuli or carnelian, were in use in Mesopotamia and Syria from the Early Bronze Age onwards.[10] Fine examples include melon-shaped pinheads made from lapis lazuli which were deposited in the Royal Cemetery of Ur (Early Dynastic III period).[11] Besides pinheads made of semiprecious stones, some evidence for the use of globular pinheads made of faience is known from the Early Dynastic period of Mesopotamia, though they are sparsely found or limited to individual sites (see, for example, the evidence from the Royal Cemetery of Ur, where only two Royal Tombs yielded faience, and only 10 out of 650 private tombs assigned to the Early Dynastic III period contained faience beads or pinheads).[12] Further evidence for the early use of composite

Figure 1. Loop pin from Enkomi Br. T. 19. Image AN46789001001 © The Trustees of the British Museum.

---

[3] Goring 1983: I, 285–288. Cf. Åström and Åström 1972: 500, type II, 1a, 2a–b ('toggle pin with a ring attached to the shaft'), 568.
[4] Åström and Åström 1972: 500, type II, 2a–b ('toggle pin with a shaft made of a so-called double loop-in-loop chain') referring to Higgins 1980: 15–16, figs. 3c–d (multiple loop-in-loop chain).
[5] Higgins 1980: 15–16, figs. 3c–d (including clues for earlier examples stemming from Ur). Further examples are given in Aldred 1990: 98–99 (gold chains dating from the 12th Dynasty – the Ramesside Period) and Demakopolou 2006: 103, pl. 33 (multiple gold chain from Mycenae; cf. Sakellariou-Xenaki 1985: 172, pl. 70).
[6] Goring 1983: II, 185–187, nos. 603–606.
[7] Br. T. 19: Murray, Smith and Walters 1900: 43, pl. VIII; Marshall 1969: 32–33, no. 550; Goring 1983: II, 186, no. 604; Crewe, Catling and Kiely 2009: no. 19.90 (max.l. 13.2cm). Br. T. 92: Murray et al. 1900: 19, fig. 38; Marshall 1969: 32, no. 549, pl. IV; Goring 1983: II, 185, no. 603, pl. 27; Crewe, Catling and Kiely 2009: no. 92.30 (max. l. 6.6cm).
[8] Gjerstad, Lindros, Sjöqvist and Westholm 1934: pl. CXLV, 4; Pierides 1971: 22–23, no. 11, pl. XI; Goring 1983: II, 186–187, no. 606, pl. 28 (max. l. 14cm).

[9] Marshall 1969: 33, no. 551; Goring 1983: II, 186, no. 605; Crewe, Catling and Kiely 2009: no. 93.40 (max. l. 9.9cm).
[10] Klein 1992: 193–222, 271.
[11] See Tallis 2011: 51, no. 26 for a recent comment on this pin. Similar pinheads made of lapis lazuli are part of the Tôd Treasure in Egypt, c. 1991–1885 BC; see Aruz, Benzel and Evans 2008: 69 (brief mention) and Quenet, Pierrat-Bonnefois, Danrey, Donnat and Lacambre 2013: 516 and 523, pl. 4.
[12] See Moorey 1999: 173 for further references.

pins with pinheads of semiprecious stone is reported from Alacahöyük in north-central Anatolia (Early Bronze Age III). This also applies to Boğazköy in central Anatolia, where several star-shaped pinheads made of bronze with traces of a frit-like substance in their grooves appeared during the Middle Bronze Age period. Initially, these Anatolian pins may have been of similar appearance to melon-shaped pinheads made of faience, such as those known from the Near East.[13]

The series of Late Cypriot loop pins, however, has been inspired by the Levantine toggle pins that often consist of a wire loop which was inserted through their eyelets. Pins of this type were part of the Byblos Treasure (c. 2000 BC), but were also discovered in Middle Bronze Age layers at Ras Shamra and Jericho, in Stratum XII at Megiddo (Middle Bronze Age), and at Tell el-Ajjul (Middle Bronze Age-Late Bronze Age).[14] Of those, two gold toggle pins from Tell el-Ajjul should be highlighted because of their globular pinheads of coloured materials: one is made of amethyst, whilst the other one is now missing (Middle Bronze Age II-Late Bronze Age).[15] Both pins were cast in one piece. The upper part of the shaft is decorated with eight or nine conical beads with horizontal incisions. More incisions on the lower part beneath the eyelet are executed in a regular pattern.

A further comparison of the latter toggle pins from Tell el-Ajjul with two more loop pins found at Enkomi illustrates the similarity between Levantine and Cypriot composite pins. In this context, it is worth noting that the Cypriot loop pins used for comparison differ from the specimens mentioned above, by virtue of the ornaments that were attached to the pin shaft. The comparison of Levantine and Cypriot pin shaft ornaments indicates that Cypriot loop pins can be understood as simplified versions of Levantine toggle pins;[16] this applies in particular to the toggle pins from Tell el-Ajjul, which offer the best parallels. The first loop pin compared here was found in Br. T. 67 at Enkomi. Unlike the second specimen discussed here, all parts of the loop pin – including a melon-shaped bead attached to the top – are made of gold (LCIA–IIC).[17] The shaft ornaments are in the form of slightly rounded beads or segments; some are fluted. Accordingly, this jewellery item imitates the design of the more elaborately executed Levantine toggle pins, such as those found at Tell el-Ajjul. The second loop pin which is used for comparison was discovered in Br. T. 19 at Enkomi (LC IA–IIC; LCIIA?).[18] Here, a turquoise faience bead was fastened to the top of the loop pin. In addition, a series of simple gold ornaments were attached to the pin shaft in an alternating pattern. The gold ornaments consist of six gold discs with fluted edges and six cylindrical beads made of twisted wire. Although the discs and beads are not precisely paralleled in the specimens available for comparison, this loop pin seems, to some extent, to reflect the design of Levantine composite pins. Furthermore, the unusual combination of gold discs and beads evokes the shape of a small hand-held distaff, a tool used for processing flax or animal fibres before spinning. Thus, the gold discs may reflect the disc-shaped mountings of the original tool which ensured that the flax fibres were held in place during the work process. Interestingly, L. and P. Åström have already established similar associations between Cypriot loop pins and distaffs which could have inspired their manufacture. Both scholars, however, prefer to assign another related type of loop pins to these tools (Figure 2).[19]

Figure 2. Loop pin from Enkomi Br. T. 19. Image AN81529001001 © The Trustees of the British Museum.

---

[13] Pieniążek and Kozal 2014: 187–208, fig. 4.3c referring to Klein 1992: 193–222.
[14] Åström and Åström 1972: 568, n. 5–7; Goring 1983: I, 294 following Chéhab 1937: pl. V, 25–26; Schaeffer 1936: 129, fig. 17; Henschel-Simon 1938: 187, fig. 11; Loud 1948: pl. 220.28; Petrie 1932: pl. III.13; and Petrie 1934: pls. XIV.20, 22–24; XV; XVI.57; XVII; XVIII.121; XIX; XX.138; XXII.251.
[15] Negbi 1971: 50, nos. 302, 308; Petrie 1934: 5–9, pl. XIV.24; pl. XVII; pl. XVIII.95.
[16] Åström and Åström 1972: 568; Goring 1983: I, 294.
[17] Murray, Smith and Walters 1900: 19, fig. 38; Marshall 1969: 33, no. 55; Goring 1983: II, 185, no. 601; Crewe, Catling and Kiely 2009: no. 67.22 (max. l. 5.95cm).
[18] Murray, Smith and Walters 1900: 43, pl. VIII; Marshall 1969: 34, no. 562; Åström and Åström 1972: 500, type II, 1; Goring 1983: II, 185, no. 602; Crewe, Catling and Kiely 2009: no. 19.91 (max. l. 8.9cm).
[19] Åström and Åström 1972: 500, type II, 1b ('toggle pin with a distaff head'; i.e., a toggle pin with a pinhead in the form of a seven-petalled flower).

As can be seen in the discussion above, the manufacture of gold and faience Cypriot loop pins was influenced by Near Eastern (Levantine) composite pins. As the use of coloured materials for the manufacture of composite pins dates back as early as the 3rd millennium BC, one might expect that pinheads made of faience were widely used within the Eastern Mediterranean. Archaeological evidence, however, is rather sparse or is limited to particular sites; this is partly due to the fact that the beads that were initially attached to the pins were destroyed or dispersed at some point in the past. However, Cypriot loop pins show some striking parallels with Levantine toggle pins. This particularly applies to two Cypriot loop pins with beaded pin shafts that are best paralleled in the toggle pins from Tell el-Ajjul. Therefore, Levantine influence on the later manufacture of loop pins on Cyprus is highly probable.[20] There can be no doubt, however, that the loop pins unearthed in Late Bronze Age tombs at Enkomi are purely Cypriot creations. In this context, attention should be drawn to the special characteristics of Cypriot loop pins, such as missing bead caps,[21] incised lines that were drawn freehand, and irregularly arranged or formed ornaments. Furthermore, an extra wire loop was attached to the pins and recalls the eyelets of the Levantine toggle pins that may have inspired the manufacture of loop pins on Cyprus.

Composite Jewellery Items II. Dress Pins combining Metal and Organic Materials

This section deals with the Late Cypriot metal pins with pomegranate-shaped heads from Enkomi and Kourion (Bamboula and Kaloriziki). This type of Cypriot composite pin is assembled from two different parts: (1) the pin shaft made of bronze or iron; and (2) the pinhead carved from ivory, bone or some similar animal material (antler).[22] The shape of the pinhead is generally associated with pomegranates or poppy seed capsules. In this context, Ch. Ward is right to point out that such objects do not represent ripe pomegranates (that are of almost globular shape) but illustrate different stages of fruit development.[23] Due to the corrosion of non-precious metals, nearly all of the pins in this series were found in fragments or in the form of pinheads of various shape and size. This means that the originally intended use of the composite pins remains difficult to determine. In addition, there is a large series of related ivory artefacts that have been variously referred to by different authors as spinning tools (spindles or distaffs), toilet articles (kohl sticks), or ceremonial sceptres, which further complicates interpretation.[24] In contrast to the pins under examination, those artefacts are assembled from an ivory rod and a detachable pomegranate-shaped finial. An interpretation problem arises, however, when only the ivory pomegranate has been preserved.

The following examples serve to illustrate the problems that arise from investigating Cypriot composite pins made of bronze/iron and ivory. These pins, which can be assigned to three different types, were discovered in Late Cypriot tombs at Enkomi and Kourion (Bamboula and Kaloriziki). The first example was selected from the funerary assemblage of Br. T. 74 at Enkomi (LC III).[25] In the research literature, this artefact is referred to as composite pin made of iron and ivory or, alternatively, as small weapon. The pin shaft is very corroded. However, since the size and the shape of the roughly globular head are best paralleled in handles of engraving tools from Enkomi, one may assume that this artefact was also an engraving tool.[26] The same may apply to a similar pinhead made of bone or antler which shows traces of an iron rod riveted onto the bottom side; this pinhead was also found at Enkomi (uncertain tomb; LC III; Figure 3).[27] The pin which was found in fragments in Tomb 19 at Kourion-Bamboula provides another example of the composite pins found on Cyprus (LC IIIB).[28] The iron pin is corroded and broken; a part of the shaft is missing. Similarly, the ivory pinhead is broken into two halves and is only partially preserved. In contrast to the previous examples found at Enkomi, the pinhead is rather small and of biconical shape. A second pin of this type, coming from Tomb 25 at Kourion-Kaloriziki, was assembled from bronze and ivory (LC IIIB).[29]

The last example is a finely carved ivory pinhead that was discovered at Enkomi (uncertain tomb; LIIC–LCIIIA?).[30] The shape of the pomegranate evokes the six-petalled calyx of the maturing pomegranate fruit. In addition to this pinhead, the Late Cypriot tombs at Enkomi yielded a second, slightly taller ivory pinhead

---

[20] Further links between Levantine toggle pins and Cypriot eyelet pins, which are considered Cypriot versions of toggle pins and form the largest group of Late Cypriot dress pins, make this all the more probable. Cf. Goring 1983: I, 298; II, 176–183, nos. 575–598a.
[21] As can be seen from the above examples of jewellery (loop pins from Enkomi), the beads of the pinheads are generally mounted with rivets whose top end was partly split into two prongs in order to hold the beads in place.
[22] Åström and Åström 1972: 474, type I ('plain pin with ivory head'), 559.
[23] Ward 2003: 538.
[24] Åström and Åström 1972: 550, types I–VI, 610 (pins or khol sticks); Kourou 1994: 207 and Ward 2003: 538 (cult objects); Sauvage 2012 and 2014 (spindles and distaffs).
[25] Murray, Smith and Walters 1900: 53; Åström and Åström 1972: 474, type I, 559; Crewe, Catling and Kiely 2009: no. 74.4: max. l. 11.2cm; diam. 3.3cm (ivory head).
[26] For details on these engraving tools see Courtois 1984: 24–25, no. 198, pl. 7.3.
[27] Crewe, Catling and Kiely 2009: no. U.1: max. l. c. 5cm; diam. 3cm (bone head). Another iron pin that is comparable with the earlier examples from Enkomi was found at Lapithos-Kastros. Its ivory head, however, was used in the turned position and could not have served as a tool handle (CG II; Gjerstad, Lindros, Sjöqvist and Westholm 1934: 215, pl. CLIII.3).
[28] Benson 1972: 131, no. 6, pl. 35: diam. c. 2.5cm (ivory head).
[29] Benson 1973: 34, no. K 1062, 122, pl. 40.
[30] Crewe, Catling and Kiely 2009: no. U.198: h. 3.4cm; diam. 2cm (head); 0.8cm (perforation).

Figure 3. Composite artefact (pin?) from Enkomi O.T. 74. Image AN107732001001 © The Trustees of the British Museum.

Figure 4. Ivory pomegranate from Enkomi (uncertain tomb). Image AN95238001001 © The Trustees of the British Museum.

which is similar, but slimmer in body than the first one. The second pinhead was found in fragments (uncertain tomb; LIIC–LC IIIA?).[31] One of these pinheads (either the first or the second) is mentioned in the excavation report and can be assigned to Br. T. 19. Interestingly, the first-mentioned ivory pinhead from Enkomi, which has remained largely unpublished, is best paralleled in an ivory pomegranate of unknown provenance that presumably resulted from illicit excavations undertaken in the Jerusalem district. The Jerusalem ivory pomegranate was first studied by A. Lemaire and acquired by the Israel Museum in 1989.[32] Since then, the authenticity of this ivory artefact – whose short paleo-Hebrew inscription may be a modern forgery – has been a matter of controversial debate.[33] In this context, it is important to recall that the ivory pomegranate is often referred to as the only existing remnant of King Salomon's temple in Jerusalem. More specifically, the Jerusalem ivory pomegranate has been interpreted as the head of an ivory sceptre or as a decorative element of the temple's cultic inventory (Figure 4).[34]

Here, some typological aspects will be taken into consideration before turning back to the interpretation of ivory pomegranates. As indicated above, there are some striking parallels between the ivory pomegranates from Enkomi and Jerusalem: (1) both ivory artefacts are worked in hippopotamus ivory; (2) the ivory artefacts are very similar in terms of shape and state of preservation (cf. the elongated body and the tall, narrow neck that is terminated by a cluster of six petals [now partially broken]); and (3) a cylindrical hole was cut into the base (c. 8–6.5mm in diameter). The hole of the Enkomi ivory pomegranate, however, was never completed, which could indicate local manufacture. In both cases, no traces of metal pin shafts that would have been inserted in the hole at the bottom can be determined, which further complicates the interpretation. Furthermore, as can be seen from the examples mentioned above, the parallels with composite pins assembled from

---

[31] Murray, Smith and Walters 1900: 14–15, fig. 24; Crewe, Catling and Kiely 2009: no. U.233.
[32] Lemaire 1981; see also Lemaire 1984.
[33] See more recently Goren, Ahituv, Ayalon, Bar-Matthews, Dahari, Dayagi-Mendels, Demsky and Lavin 2005 and Ahituv, Demsky, Goren and Lemaire 2007. The inscription is translated as 'Belonging to the Temp[le of Yahw]eh, holy to the priests' (Lemaire 1984) or as 'Sacred donation for the priests of [in]the House of Yahweh' (Avigad 1990: esp. 160).
[34] See for instance Avigad 1990: 162–166.

bronze/iron and ivory are not close enough to suggest that ivory pomegranates, such as the specimens from Enkomi, served as pinheads. Therefore, typological comparisons are limited to a series of related ivory artefacts that includes ivory rods with pomegranate-shaped heads of similar (but not identical) design. This can be demonstrated by examining a pair of ivory rods found in Sw. T. 3 at Enkomi (LCIB-IIC).[35] The same applies, though to a lesser degree, to an ivory rod with pomegranate head that was found in Tomb 9 at Kition (upper burial, c. 1200 BC).[36] Additionally, three ivory pomegranates were found in the Kition temples. These artefacts, however, differ in size and are also referred to as poppy capsules (LCIIC/IIIA).[37] Based on these considerations, finely carved ivory pomegranates, such as the two specimens from Enkomi, can no longer be assigned to the class of composite pins. Instead, such artefacts may instead belong to the class of ivory rods with decorated finials that could have served as spinning tools, toilet articles or ceremonial sceptres. In this respect, mention should be made of more recent surveys on the evidence stemming from the Levant (Ugarit, Megiddo, Lachish, Tell Dan, Hama, and Tell Kazel) and from elsewhere (Perati, Delos, Uluburun shipwreck).[38] However, further research is necessary to comprehensively clarify the interpretation of Late Bronze Age ivory pomegranates and indication of local manufacture.

Composite Jewellery Items III. The Gold Necklace from Arsos

The class of composite jewellery items from Prehistoric, Protohistoric and Archaic Cyprus also includes an outstanding example of Cypriot gold work. In this context, attention should be drawn to a gold necklace from Arsos which was found in the sanctuary of Aphrodite Golgeia in 1910 (CA I).[39] The necklace is assembled from two different parts: (1), a series of 39 (originally 40) finely granulated gold beads whose joints are covered by fine granulation; (2), a cylindrical, gold-mounted agate pendant. The gold caps of the pendants are decorated with small pyramids of gold granules. The pendant itself is richly decorated with a gold bee flanked by two uraei wearing the crowns of Upper and Lower Egypt. The necklace is a prime example of Cypriot gold work and reflects both Aegean and Near Eastern goldsmithing traditions. On the one hand, granulated gold beads have a long tradition and are well known from Mycenaean Greece and Crete.[40] With regard to contemporaneous jewellery items, mention should be made here of a granulated gold bee pendant from the base of the statue of Artemis of Ephesus (Artemision, 7th century BC).[41] Two granulated bees are part of a gold diadem with a griffin head that comes from Melos (second half of the 7th century BC).[42] Near Eastern links have been indicated in a recent article written by P. Flourentzos, who refers to the bee as important element of Egyptian symbolism.[43] Here, further links will be established with Phoenician jewellery from Carthage that is fashioned in Egyptianising style (7th/6th centuries BC). More specifically, the granulated decoration of two gold pendants should be mentioned, which can help us to understand the iconography of the composite pendant from Arsos. The discoid or niche-shaped pendants are part of a larger series of similar jewellery items. The winged sun is represented at the top of the pendants. Two uraei that are wearing the Egyptian double crown are represented at the bottom. In one case they are flanking a dome-shaped *sacrum*. In the other, they are represented on a shrine or a small temple (mid 7th–6th century BC).[44] Returning to the composite pendant that is part of the gold necklace from Arsos, it can be suggested that the agate pendant reflects a similar representation which was known to the Cypriot goldsmiths. In this context, the agate pendant can be interpreted as a (Phoenician) *sacrum*, flanked by two ureai that serve as a support for these snakes. Similarly, the two-dimensional representation of the winged sun was apparently replaced by the three-dimensional representation of a bee (Figures 5-6).

Jewellery with Colourful Designs. The Broad Collar from Enkomi

The investigation of colourful jewellery from Prehistoric and Protohistoric Cyprus also includes a masterly executed gold collar that was part of the funerary assemblage of Br. T. 93 at Enkomi (Egyptian work; 18th Dynasty).[45] Surprisingly, this outstanding piece of jewellery has so far undergone very little study. The collar consists of several rows of different pendants made of inlaid gold which were arranged so as to create a large pectoral, covering neck and shoulders. Collars of this type are well known from ancient Egypt, which has yielded numerous examples made of various coloured materials, such semi-precious stones and faience. Additionally, these collars are also represented in Egyptian figurative art.[46] They are called *usekh* collars

---

[35] Gjerstad, Lindros, Sjöqvist and Westholm 1934: pl. LXXVIII, nos. 240-241.
[36] Karageorghis 1974: 69, no. 132, 91, fig. CLVV, pl. LXXVII.
[37] Karageorghis and Demas 1985: 248, pl. CXCI.
[38] Cf. Gachet-Bizollon 2007; Sauvage 2012; Sauvage 2014 (*non vidi*).
[39] Gjerstad, Lindros, Sjöqvist and Westholm 1937: 597-598, pl. CCV.3-4; Dikaios 1953: pl. XXVII.3; Pierides 1971: 27, pl. XV.1-3. For this type of jewellery see Laffineur 1991: 171-180.
[40] Higgins 1980: 73-74, fig. 13.
[41] Becatti 1955: 164, nos. 159a-b.
[42] Despini 1996: 60, 212, nos. 17-18, pls. 27-28.
[43] Flourentzos 2009.
[44] Quillard 2014: 206-207, nos. 87-88, pls. 87-88.
[45] Murray, Smith and Walters 1900: 41, pl. V; Marshall 1969: 36-37, no. 581, fig. 6, pl. V; Åström and Åström 1972: 505 ('pectoral'), 578-579; Goring 1983: I, 254 and II, 168-169, no. 559, pl. 23; Crewe, Catling and Kiely 2009: no. 93.194.
[46] See or instance Davies 1925: pl. XI (wall paintings of the tomb of Nebamun and Ipuki at Thebes, c. 1370 BC).

Figure 5. Necklace from the sanctuary at Arsos (detail). Photo by the author. Courtesy of the Department of Antiquities, Cyprus.

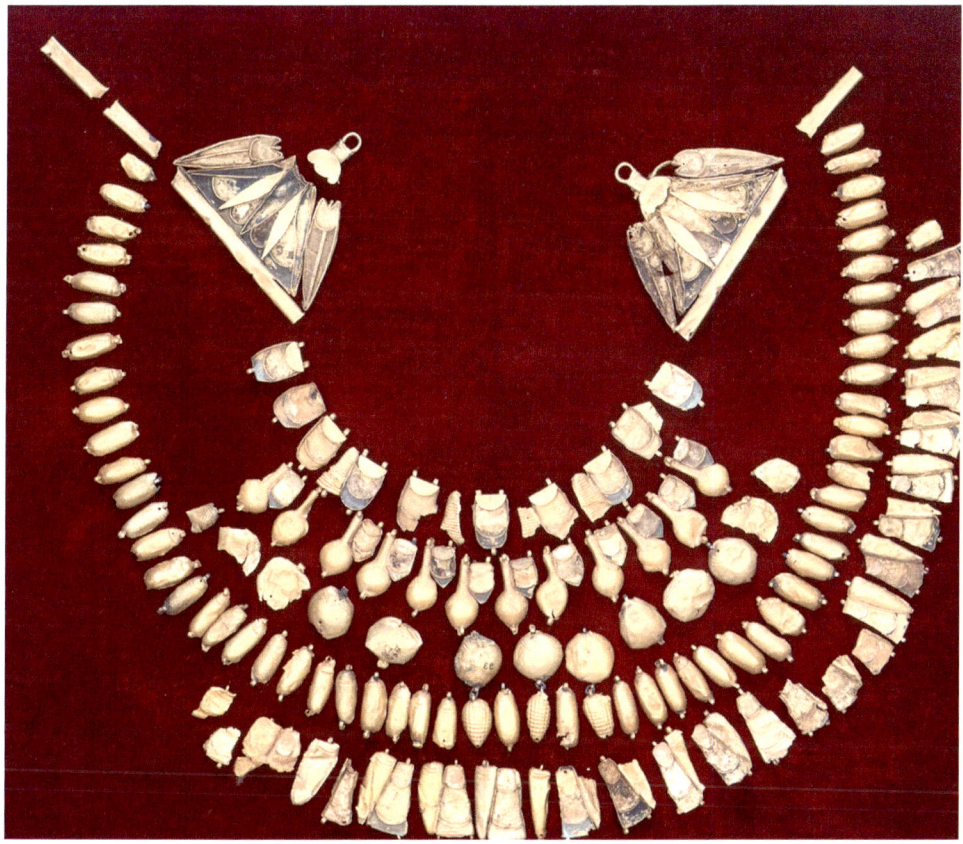

Figure 6. Broad collar (*usekh* collar) from Enkomi Br. T. 93. Image AN51096001001 © The Trustees of the British Museum.

and were fastened with large clasps. The collar from Enkomi was assembled from at least seven different types of pendants:

- sixty-two elongated leaf-shaped pendants with two depressed lines across one end (type 1);
- twenty-one double pendants which are leaf-shaped or roughly triangular (type 2);
- ten double pendants which are semi-elliptical or bell-shaped (type 3);
- 11 double pendants which are shield-shaped or represent the *nefer*-hieroglyph (type 4). This

series includes a fragment of a twelfth pendant of this type;
- 13 convex discs (type 5);
- three acron-shaped pendants (type 6);
- two large gold claps in the form of lotus flowers (type 7). In addition, four fragments were found that resulted from shoulder-pieces or gold claps.

Four series of pendants (types 2–3; type 4 [shield-shaped pendants]; type 7) are divided into cloisons filled with colourful material: blue, white, and red (Figures 7-8).

The Enkomi collar is best paralleled in the *usekh* collar from Tomb KV 55 at Thebes (18th Dynasty/Amarna period).[47] Today, the collar is assembled from five rows of different pendants. However, since each shoulder-piece is perforated with a sixth string hole, it is evident that some gold pendants are missing. The surviving pendants are similar to those of the Enkomi *usekh* collar. This applies above all to 18 double pendants which are semi-elliptical or bell-shaped and horizontally divided into cloisons filled with coloured (green and red) material. The series that follow are composed of 36 *nefer*-shaped pendants[48] and of 36 leaf-shaped pendants. These series are followed by two rows of 36 pendants of mixed types (drop-shaped, *nefer*-shaped, and leaf-shaped pendants). The collar was found with the body of the mummy in Tomb KV 55 (King's Valley no. 55 or 'Amarna Cache'), but there are no records concerning the exact find place or the assemblage of pendants.[49] The mummy of this tomb was first assigned to Queen Tiyi and later to Smenkhare and Akhenaton Smenkhare and Akhenaton.[50]

Another similar collar was found in the so-called Tomb of the Three Foreign Wives of Thutmosis III, Wady Gabbanat el-Qurud (c. 1504–1450 BC).[51] Again, this collar was reconstructed in modern times. Today, it is assembled from five rows of *nefer*-shaped pendants that are combined with a single row of palmette-shaped pendants; these pendants alternate with tiny spherical beads. Initially, the pendants were inlaid with carnelian, turquoise and blue frit. Now, the inlays are, for the most part, missing. In the research literature, two or

Figures 7-8. Finger ring from Kouklia-Evreti T. VIII (upper surface/profile). Photos by the author. Courtesy of the Department of Antiquities, Cyprus.

more different versions of re-stringing can be found. In earlier interpretations, the collar was completed by decorative clasps on which the name of Thutmosis III is mentioned.[52] These clasps have been left aside in the more recent research literature.[53] Due to the context of the tomb, however, this *usekh* collar can be assigned to Menhet, Menwi and Merti, three foreign-born wives of non-Egyptian (Syrian?) origin of Thutmosis III. The last collar which will be used for comparison has only been preserved in fragmentary form (18th–19th Dynasties).[54] The jewellery, purportedly from Thebes, consists of a single gold clasp and six short rows of gold pendants. Four out of six different types of pendants are comparable with the pendants of types 1–4 that were used for the manufacture of the *usekh* collar found at Enkomi.

Before concluding this section, it is worth noting that two of the three *usekh* collars which offer the best

---

[47] Davis 1910: pl. XXI; Aldred 1971: 211, pl. 71; Forbes 1998: 278.
[48] The Egyptian hieroglyph *nefer* stood for 'good/beauty' and was part of female personal names (e.g. Nefertiti; Nefertari). The individual sign also symbolized conceptions, such as happiness or good fortune. Contrary to the findings of older research literature, this hieroglyph is no longer considered as a lute, but associated with a (sheep's) heart and the trachea, cf. Wilkinson 1992: 79. In Egyptian jewellery, the hieroglyph is used in stylized form, as amulet or pendant.
[49] Forbes 1998: 249-309.
[50] For a more recent debate on the mummy in Tomb KV 55, see Reeves 2005 (Akhenaton). From September 2007–October 2009, eleven royal mummies underwent detailed scientific examination. This also applied to the mummy found in Tomb KV 55 that was identified as the father of Tutankhamen (Akhenaton), cf. Hawass, Gad, Ismal and Khairat 2010.
[51] Winlock 1948: 20, pl. X; Aldred 1971: 208-209, pl. 66; Lilyquist 2003: 169-170, 173, no. 3, pl. 164.

[52] Winlock 1948: pl. X; Aldred 1971: pl. 66.
[53] See for instance Müller and Thiem 2006: pl. 348.
[54] Aldred 1971: pl. 60.

parallels for the broad collar from Enkomi can be attributed to high-ranking persons, such as an Egyptian pharaoh or (foreign) women that were given the title of the King's Wife.[55] This aspect is all the more important since there is – apart from the gold sceptre from Kourion[56] – a general lack Prehistoric or Protohistoric artefacts that can be linked to the Royal Court, or that could be classified as royal insignia.

Materials and Colours Used in Cypriot Jewellery

The following section will comment on the materials and colours used in Prehistoric and Protohistoric jewellery. As demonstrated above, there are two different classes of Cypriot jewellery: (1) composite jewellery items assembled from different metals (gold, bronze or iron) and of coloured material (faience, ivory, semi-precious stone); and (2) jewellery with colourful designs, i.e., jewellery inlaid with coloured materials.

First of all, it should be noted that no distinction is made here between gold and electrum (the natural alloy of gold and silver). This is due to a lack of information in research literature, which still contains only a few scientific data relating to the metals used by ancient goldsmiths. Secondly, it should be recalled here that dress pins made of non-precious metals, such as bronze or iron, were initially free from corrosion and looked like gold or silver jewellery. Thirdly, a short definition of ancient faience should be added. H.C. Beck, who established the classification and nomination of Egyptian beads and pendants, has already suggested that the term 'faience' should be replaced by 'glazed composition'.[57] In research literature, both terms are used to designate colourful beads from the ancient Near East that consist of 'glazed frit', i.e., 'of a highly siliceous body coated with glaze...that is generally coloured, often being blue or green, though it may be of any colour'.[58]

Another important issue is the first instance of the application of enamel in antiquity. In view of numerous examples of jewellery with colourful designs from ancient Egypt, the question was raised whether enamel was used by Egyptian goldsmiths. So far, most potential evidence for enamelling was rejected. The same is true for the *usekh* collar found at Enkomi: following L. Crewe, H. Catling, and Th. Kiely, the inlaid material is not enamel but blue, red, and white vitreous paste.[59] J. Ogden, however, is thought to have identified true enamel applied to an Egyptian gold bowl found in the tomb of Undjebauendjet at Tanis (Third Intermediate or Late Period).[60] A series of six enamelled gold finger rings that were found in Tomb VIII at Kouklia-Evreti bear the same decoration and can be used to support this hypothesis.[61] Apart from these finger rings, it is known that the gold sceptre from Kourion is partly covered with true enamel.[62] Despite this wide range of materials used in Cypriot jewellery, the coloured materials were not used to create naturalistic representations of flora and fauna. Instead, the focus was on the application of rare materials and techniques in order to reflect the wealth and social status of the owner. This includes the enjoyment of expressive colour combinations and patterns, an objective that was best achieved through the application of inlay. Inlay is often but incorrectly referred to as cloisonné technique (enamel or enamelling). Instead of true enamel, coloured material such as white, blue or red vitreous paste was used in the form of pre-processed products that were set in the cloisons without any melting process. With regard to future studies, it is worth noting that the compilation of a corpus of colourful jewellery from the ancient Near East would create a good foundation for new investigations.

Fine examples of colourful jewellery from Cyprus include a series of pins with faience beads or ivory heads that were found at Enkomi, the gold necklace from Arsos that combines gold beads and elements of semi-precious stone, and the Egyptian *usekh* collar with inlaid ornaments made of vitreous paste that was unearthed at Enkomi. Jewellery items or gold work containing true enamel, however, have been found only in a small number. This may be due to the fact that enamel can easily be damaged during use, but hardly be repaired: reheating of enamelled jewellery will damage the surface even further. Thus, these jewellery items are, so to speak, deprived of recycling or can only be reused after reworking (e.g. inlay instead of enamel).

Conclusions

In this article, selected examples of colourful jewellery from Prehistoric, Protohistoric and Archaic Cyprus which have so far been little studied were discussed. As a result of this investigation, it was shown that not only the coloured materials, but also the jewellery itself, have strong links with Near Eastern art. As can be demonstrated using composite pins, one may assume that hastily or less meticulously made jewellery could be a Cypriot response to Near Eastern art. Furthermore, certain composite pins assembled from bronze/iron and ivory/bone may have been used as tools rather than as personal ornaments. Similarly, the interpretation of the small ivory pomegranates that are referred to as

---

[55] Although the latter may have been of minor rank compared to that of the Great Royal Wife.
[56] McFadden 1954. Cf. Karageorghis 2002: 135, fig. 35 and Matthäus and Matthäus-Schumacher 2012: 57–59.
[57] Beck 1929: 35.
[58] Bosse-Griffiths 2001: 47–48 referring to Lukas 1932: 61.
[59] Crewe, Catling and Kiely 2009: no. 93.194.

[60] Ogden 1990; Goring 1995: 107–109; Ogden 2000: 165–166.
[61] Catling 1968: 162–169; Pini 2010: 40–43, nos. 64–69, pls. XXXIa-XXXIVb and XVIIIc–XXIa (coloured plates).
[62] Cf. n. 56 (supra).

remains of spinning tools, toilet articles or ceremonial sceptres still needs clarification. Surprisingly, a Cypriot ivory pomegranate that is classified as a pinhead is best paralleled in an ivory pomegranate that is believed to be the only cult object remaining from King Salomon's temple in Jerusalem. This hypothesis, however, is a matter of ongoing debate. The present study also included two masterly executed jewellery items: a gold necklace from the sanctuary of Aphrodite Golgeia at Arsos, and an Egyptian *usekh* collar that was discovered at Enkomi. The finely granulated gold necklace from Arsos is as yet unparalleled in Cypriot and Near Eastern art. However, the granulated gold bee that is mounted at the top of the agate pendant finds some parallels in contemporaneous jewellery from Melos and Ephesus (Artemision). Furthermore, close iconographic links can be established with Phoenician jewellery from Carthage. In turn, the *usekh* collar found at Enkomi is best paralleled in two *usekh* collars from ancient Egypt which can be assigned to individual persons, such as the mummy found in Tomb KV 55 (Pharaoh Akhenaton?) or the three foreign wives of Thutmosis III. This is all the more important since Cyprus has only provided a few personal data and evidence of the ruling élite.

However, some questions arise from this investigation. First is the question of whether jewellery could reflect cultural identity and choices, i.e., whether it could be considered as indicative of cross-cultural contacts. The extent of influence of Near Eastern goldsmithing traditions is a further point of discussion. In addition to this, the first application of the cloisonné technique and enamel in the Eastern Mediterranean are research issues that still need clarification. Touching on all these points, this article aims to contribute to further investigations in this area of research.

Acknowledgements

I would like to thank Dr. Silvana di Paolo, Instituto di Studi sul Mediterraneo Antico–CNR (Rome), who kindly invited me to participate in her workshop on the composite artefacts in the ancient Near East and to give a paper about colourful jewellery from ancient Cyprus.

References

Ahituv, S., A. Demsky, Y. Goren and A. Lemaire 2007. The Inscribed Pomegranate from the Israel Museum Examined Again, *Israel Exploration Journal* 57/1: 87–95.

Aldred, C. 1971. *Jewels of the Pharaohs. Egyptian Jewellery from the Dynastic Period*. London: Thames & Hudson.

Aruz, J., K. Benzel and J.M. Evans (eds) 2008. *Beyond Babylon. Art, Trade, and Diplomacy in the Second Millennium BC*. New York: Metropolitan Museum of Art.

Åström, L. and P. Åström 1972. *The Late Cypriote Bronze Age. Other Arts and Crafts. The Swedish Cyprus Expedition Vol. IV.1D*. Lund: Swedish Cyprus Expedition.

Avigad, N. 1990. The Inscribed Pomegranate from the 'House of the Lord', *Biblical Archaeologist* 53: 157–166.

Becatti, G. 1955. *Oreficerie antiche dalle minoiche alle barbariche*. Rome: Istituto Poligrafico dello Stato.

Beck, H.C. 1929. Classification and Nomination of Beads and Pendants, *Archaeologia* 77: 1–76.

Benson, J.L. 1972. *Bamboula at Kourion. The Necropolis and the Finds*. Philadelphia: University of Pennsylvania Press.

Benson, J.L. 1973. *The Necropolis of Kaloriziki: Excavated by J.F. Daniel and G.H. McFadden for the University Museum, Pennsylvania, Philadelphia*, Studies in Mediterranean Archaeology 36. Göteborg: P. Åström Forlag.

Bombardieri, L., A. D.'Agostino, G. Guarducci, V. Orsi and S. Valentini (eds) 2013. *SOMA 2012. Identity and Connectivity. Proceedings of the 16th Symposium on Mediterranean Archaeology, Florence, Italy, 1-3 March 2012*, BAR-International Series 2581, Vol. 1. Oxford: Archaeopress.

Bosse-Griffiths, K. 2001. *Amarna Studies and Other Selected Papers*. Fribourg: University Press.

Catling, H.W. 1968. Kouklia: Evreti Tomb 8, *Bulletin de Correspondance Hellénique* 92: 162–169.

Chéhab, M. 1937. Un trésor d'orfèvrerie syro-égyptien, *Bulletin du Musée de Beyrouth* I: 7–21.

Courtois, J.-C. 1984. *Alasia III. Les objets des niveaux stratifiés d'Enkomi. Fouilles C.F.A. Schaeffer (1947–1970)*. Paris: Éditions Recherche sur les civilisations.

Crewe, L., H. Catling and Th. Kiely 2009. Enkomi. In *Ancient Cyprus in the British Museum*. Online Research Catalogue. www.britishmuseum.org/research/publications/online_research_catalogues/ancient_cyprus_british_museum.aspx

Dalton, M. 2007. The Egkomi Mapping Project, *Report of the Department of Antiquities, Cyprus*: 157–174.

Davies, N. de G. 1925. *The Tomb of Two Sculptors at Thebes*. New York: Metropolitan Museum of Art.

Davis, Th.M. 1910. *The Tomb of Queen Tiyi*. London: Constable.

Demakopoulou, K. (ed.) 2006. *The Aidonia Treasure: Seals, and Jewellery of the Aegean Late Bronze Age, Athens, National Archaeological Museum, 30 May-1 September 1996*. Athens: Ministry of Culture, Archaeological Receipts Fund.

Despini, A. 1996. *Greek Art. Ancient Gold Jewellery*. Athens: Ekdotiki Athinon.

Dikaios, P. 1953. *A Guide to the Cyprus Museum*. Nicosia: Cyprus Government Printing Office.

Flourentzos, P. 2009. The Relations of Egyptian Iconography and Symbolism with the Royal Ideology of Cypriot City Kingdoms. In D. Michaelides, V. Kassianidou and R.S. Merrillees (eds), *Egypt and Cyprus in Antiquity. Proceedings of the International Conference, Nicosia, 3-6 April 2003*: 67–77. Oxford: Oxbow Books.

Forbes, D.C. 1998. *Tombs. Treasures. Mummies. Seven Great Discoveries of Egyptian Archaeology*. Sebastopol (CA): KMT Communications.

Gachet-Bizollon, J. 2007. *Les ivoires d'Ougarit et l'art des ivoiriers du Levant au Bronze Récent, Ras Shamra-Ougarit*, XVI. Paris: Éditions Recherche sur les Civilisations.

Gjerstad, E., J. Lindros, E. Sjöqvist and A. Westholm 1934. *The Swedish Cyprus Expedition Vol. 1. Finds and Results of the Excavations in Cyprus 1927-1931*. Stockholm: The Swedish Cyprus Expedition.

Gjerstad, E., J. Lindros, E. Sjöqvist and A. Westholm 1937. *The Swedish Cyprus Expedition Vol. 3. Finds and Results of the Excavations in Cyprus 1927-1931*. Stockholm: The Swedish Cyprus Expedition.

Goren, Y., S. Ahituv, A. Ayalon, M. Bar-Matthews, U. Dahari, M. Dayagi-Mendels, A. Demsky and N. Lavin, A Re-Examination of the Inscribed Pomegranate in the Israel Museum, *Israel Exploration Journal* 55/1: 3–20.

Goring, E.S. 1983. *Late Cypriot Goldwork* (PhD Diss., Bedford College, London).

Goring, E.S. 1995. The Kourion Sceptre: Some Facts and Factoids. In Ch. Morris (ed.), *Klados. Essays in Honour of J.N. Coldstream*: 103–110. London: Institute of Classical Studies.

Hawass, Z., Y.Z. Gad, S. Ismal and R. Khairat, Ancestry and Pathology in King Tutankhamun's Family, *Journal of American Medicine* 303/7: 638–647.

Henschel-Simon, E. 1938. The 'Toggle-Pins' in the Palestine Archaeological Museum, The *Quarterly of the Department of Antiquities in Palestine* 6: 169–209.

Higgins, R.A. 1980. *Greek and Roman Jewellery*, 2nd edition. Berkeley and Los Angeles: University of California Press.

Karageorghis, V. 1974. *Excavations at Kition. Vol. I. The Tombs*. Nicosia: Department of Antiquities.

Karageorghis, V. 2002. *Early Cyprus. Crossroads of the Mediterranean*. Los Angeles: J. Paul Getty Museum.

Karageorghis, V. and M. Demas 1985. *Excavations at Kition V. The Pre-Phoenician Levels*. Nicosia: Republic of Cyprus, Ministry of Communications and Works, Department of antiquities.

Klein, H. 1992. *Untersuchung zur Typologie bronzezeitlicher Nadeln in Mesopotamien und Syrien*, Schriften zur Vorderasiatischen Archäologie 4. Saarbrücken: Saarbrücker Druckerei und Verlag.

Kontomichali, M. 2002. *La bijouterie à Chypre aux époques chypro-géométrique et chypro-archaïque* (PhD Diss., Université de Lyon II, Lyon).

Kourou, N. 1994. Sceptres and Maces Before, During and Immediately after the 11th century BC. In V. Karageorghis (ed.), *Cyprus in the 11th century B.C.*: 203–215. Nicosia: Leventis Foundation.

Laffineur, R. 1991. La bijouterie chypriote d'après le témoignage des terres cuites: l'exemple des statuettes d'Arsos. In F. Vandenabeele and R. Laffineur (eds), *Cypriote Terracottas. Proceedings of the First International Conference of Cypriote Studies, Brussel - Liège - Amsterdam, 29 May - 1 June, 1989*: 171–181. Brussels and Liège: Leventis Foundation, Vrije Universiteit Brussel and Université de Liège.

Lemaire, A. 1981. Une inscription paléo-hébraïque sur une grenade en ivoire, *Revue Biblique* 88/2: 236–239.

Lemaire, A. 1984. Probable Head of Priestly Sceptre from Solomon's Temple Surfaces in Jerusalem, *Biblical Archaeology Review* 10/2: 24–29.

Lilyquist, Ch. 2003. *The Tomb of the Three Foreign Wives of Thutmosis III*. New York: The Metropolitan Museum of Art.

Loud, G. 1948. *Megiddo II. Seasons of 1935-1939*. Chicago: University Press.

Lucas, A. 1932. *Antiquities. Their Restoration and Preservation*, 2nd edition. London: Edward Arnold & Co.

Marshall, F.H. 1969. *Catalogue of the Jewellery, Greek, Etruscan, and Roman, in the Departments of Antiquities, British Museum*, 2nd edition. London: British Museum.

Matthäus, H. and Schumacher-Matthäus, G. 2013. Social and Ethnic Change in Cyprus during the 11th century B.C. – New evidence from Kourion, Kaloriziki, Tomb 40. In *Actes du Colloque International 'Un millénaire d'histoire et d'archéologie chypriote (1600 - 600 av. J.-C.)', Milan, 18-19 octobre 2012*, *Pasiphae* 7: 159–168.

McFadden, G.H. 1954. A Late Cypriote III Tomb from Kourion, *American Journal of Archaeology* 58: 131–141.

Michaelides, D., V. Kassianidou and R.S. Merrillees (eds) 2009. *Egypt and Cyprus in Antiquity. Proceedings of the International Conference, Nicosia, 3-6 April 2003*. Oxford: Oxbow Books.

Moorey, P.R.S. 1999. *Ancient Mesopotamian Materials and Industries. The Archaeological Evidence*. Winona Lake: Eisenbrauns.

Müller, H.W. and E. Thiem 2006. *Die Schätze der Pharaonen*. Munich: Weltbild Verlag.

Murray, A.S., A.H. Smith and H.B. Walters 1900. *Excavations in Cyprus*, 2nd edition. London: Trustees of the British Museum.

Negbi, O. 1971. The Hoards of Goldwork from Tell el-Ajjul, Studies in Mediterranean Archaeology 25. Göteborg: Paul Âströms.

Ogden, J. 1990. Gold in the Time of Bronze and Iron, *The Journal of the Ancient Chronology Forum* 4: 6–14.

Ogden, J. 2000. Metals. In P.T. Nicholson and I. Shawn (eds), *Ancient Egyptian Materials and Technology*: 148–176. Cambridge: University Press.

Petrie, W.M.F. 1932. *Ancient Gaza (Tell el-Ajjul)*, vol. 2. London: British School of Archaeology in Egypt.

Petrie, W.M.F. 1934. *Ancient Gaza (Tell el-Ajjul)*, vol. 4: London: British School of Archaeology in Egypt.

Pieniążek, M. – Kozal, E. 2014. West Anatolian Beads and Pins in the 2nd millennium BC. Some Remarks on Function and Distribution in Comparison with Neighbouring Regions, *Polish Archaeology in the Mediterranean* 23/2: 187–208.

Pierides, A. 1971. *Jewellery in the Cyprus Museum*. Nicosia: Department of Antiquities.

Pini, I. 2010. *Aegean and Cypro-Aegean Non-Sphragistic Decorated Finger Rings of the Bronze Age*, Aegaeum 31. Liège and Austin: Université de *Liège* and University of Texas at Austin.

Quenet, Ph., G. Pierrat-Bonnefois, V. Danrey, S. Donnat and D. Lacambre, New Lights on the Lapis lazuli of the Tôd Treasure, Egypt. In Bombardieri, D.'Agostino, Guarducci, Orsi and Valentini 2013: 515–525.

Quillard, B. 2014. Carthaginian Jewelry. In J. Aruz, S.B. Graff and Y. Rakic (eds), *Iberia to Syria at the Dawn of the Classical Age*: 206–210. New York: The Metropolitan Museum of Art.

Reeves, N. 2005. *Akhenaten: Egypt's False Prophet*. London: Thames & Hudson.

Sakellariou-Xenaki, A. 1985. Οἱ θαλαμωτοί τάφοι τῶν Μυκηνῶν ἀνασκαφῆς Χρ. Τσούντα (1887–1898). Paris: de Boccard.

Sauvage, C. 2012. Spinning from Old Threads: the Whorls from Ugarit at the Musée d'Archéologie Nationale of Saint-Germain-en-Laye and at the Louvre. In H. Koefoed, M.L. Nosch and E. Anderson Strand (eds), *Textile Production and Consumption in the Ancient Near East. Archaeology, Epigraphy, Iconography:* 189–214. Oxford: Oxbow Books.

Sauvage, C. 2014. Spindles and Distaffs: Late Bronze and Early Iron Age Eastern Mediterranean Use of Solid and Tapered Ivory/Bone Shafts. In M.-L. Nosch, C. Michel *and* M. Harlow (eds), *Prehistoric, Ancient Near Eastern & Aegean Textiles and Dresses. An Interdisciplinary Anthology*: 184–226. Oxford and Oakville: Oxbow Books.

Schaeffer, C.F.A. 1936. Les fouilles de Ras Shamra à Ougarit. Septième campagne (printemps 1935). Rapport Sommaire, *Syria* 17/2: 105–149.

Tallis, N. (ed.) 2011. *Splendours of Mesopotamia, Abu Dhabi, 29 March to 27 June 2011*. London: British Museum Press.

Ward, Ch. 2003. Pomegranates in the Eastern Mediterranean during the Late Bronze Age, *World Archaeology* 34/3: 529–541.

Wilkinson, R.H. 1992. *Reading Egyptian Art: a Hieroglyphic Guide to Ancient Egyptian Painting and Sculpture*. London: Thames and Hudson.

Winlock, H.E. 1948. *The Treasure of the Three Egyptian Princesses*. New York: Metropolitan Museum of Art.